ON THE MOTION OF THE HEART AND BLOOD IN ANIMALS

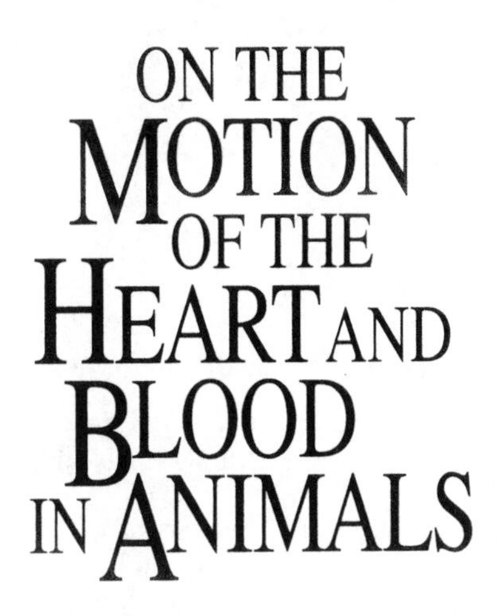

ON THE MOTION OF THE HEART AND BLOOD IN ANIMALS

WILLIAM
HARVEY

Translated by Robert Willis

GREAT MINDS SERIES

PROMETHEUS BOOKS
Amherst, New York

Published 1993 by Prometheus Books

59 John Glenn Drive, Buffalo, New York 14228-2119
716-691-0133 ext. 210 FAX: 716-691-0137

Library of Congress Cataloging-in-Publication Data

Harvey, William, 1578–1657.
[De motu cordis. English]
On the motion of the heart and blood in animals / William Harvey ; translated by Robert Willis.
p. cm. — (Great minds series)
Originally published: New York : P.F. Collier & Son, 1910.

ISBN 13: 978-087975-854-7
ISBN 10: 0-87975-854-6 (pbk.)

1. Blood—Circulation—Early works to 1800. 2. Heart—Early works to 1800. I. Title. II. Series.
[DNLM: WZ 290 H342d 1910a]
QP101.H3613 1993
612.1'3—dc20
DNLM/DLC
for Library of Congress 92-25948
CIP

Printed in the United States of America on acid-free paper.

Great Minds Series
(Classics in Life Sciences)

- ❑ Jacob Bronowski—*The Identity of Man*
- ❑ Francis Crick—*Of Molecules and Men*
- ❑ Charles Darwin—*The Autobiography of Charles Darwin*
- ❑ Charles Darwin—*The Descent of Man*
- ❑ Charles Darwin—*The Origin of Species*
- ❑ Charles Darwin—*The Voyage of the Beagle*
- ❑ René Descartes—*Treatise of Man*
- ❑ Ernst Haeckel—*The Riddle of the Universe*
- ❑ William Harvey—*On the Motion of the Heart and Blood in Animals*
- ❑ Julian Huxley—*Evolutionary Humanism*
- ❑ Thomas H. Huxley—*Evolution and Ethics* and *Science and Morals*
- ❑ Edward Jenner—*Vaccination against Smallpox*
- ❑ Louis Pasteur & Joseph Lister—*Germ Theory and Its Applications to Medicine (Pasteur)* and *On the Antiseptic Principle of the Practice of Surgery (Lister)*
- ❑ Alfred R. Wallace—*Island Life*

See the back of this volume for a complete list of titles in Prometheus's Great Books in Philosophy and Great Minds series.

WILLIAM HARVEY, the oldest son of Thomas Harvey, was born at Folkestone, Kent, England, on April 1, 1578. He was educated at King's School, Canterbury, and Cambridge University, where he received his B.A. at the age of nineteen. Having chosen medicine as his career, Harvey entered the University of Padua, where he studied under the famed anatomist Girolamo Fabrici (Hieronymus Fabricius ab Aquapendente) and his pupil Giulio Casserio.

Upon receiving his M.D. in 1602, Harvey returned to England, where he married the daughter of Dr. Lancelot Brown, former physician to Elizabeth I; he became a fellow of the Royal College of Physicians in 1607, physician to St. Bartholomew's Hospital in 1609, and Lumleian lecturer at the College of Physicians. In 1616, Harvey began a series of lectures on the movement of the heart and blood.

Appointed physician extraordinary to King James I, Harvey maintained close professional ties to the royal family until the end of the Civil War. Under Charles I, Harvey was made warden of Merton College, Oxford, in 1645, but resigned a year later due to advancing age; thereupon he was offered the presidency of the College of Physicians. William Harvey died, in his eightieth year, on June 7, 1657.

Harvey's groundbreaking work, *Exercitatio anatomica de motu cordis et sanguinis in animalibus* (*On the Motion of the Heart and Blood in Animals*), including explanations of heart valves and arterial pulse, was published in Latin in 1628. Despite increasing knowledge about blood circulation and the discovery of valves in the veins, the studies of physiology and anatomy in Harvey's day were still dominated by ancient Greek philosophy and medicine, notably Aristotle and Galen, who taught that in human beings and the higher animals, the blood was elaborated in the liver and sent to the heart, from where it circulated through the body. While the veins carried blood, the arteries carried both blood and a kind of vital air, or spirit (*pneuma*).

Harvey's work was revolutionary in that it relied extensively on minute observations of living animals as well as on a lengthy series of dissections, which gave Harvey a far more complete knowledge of the comparative anatomy of the heart and vessels than any of his contemporaries—a knowledge not to be equaled until the researches of the anatomists John Hunter and Johann Friedrich Meckel in the eighteenth and early nineteenth centuries. Although Harvey's discoveries were met with some reluctance at first, before his death his views were acknowledged and honored throughout Europe. *On the Motion of the Heart and Blood in Animals* remains one of the greatest works of physiology.

William Harvey also wrote *Exercitationes de generatione animalium* (*On Animal Reproduction*) (1651).

DEDICATION

TO HIS VERY DEAR FRIEND,
DOCTOR ARGENT, THE EXCELLENT AND
ACCOMPLISHED PRESIDENT OF THE ROYAL COLLEGE OF PHYSICIANS,
AND TO OTHER LEARNED PHYSICIANS, HIS
MOST ESTEEMED COLLEAGUES.

I HAVE already and repeatedly presented you, my learned friends, with my new views of the motion and function of the heart, in my anatomical lectures; but having now for more than nine years confirmed these views by multiplied demonstrations in your presence, illustrated them by arguments, and freed them from the objections of the most learned and skilful anatomists, I at length yield to the requests, I might say entreaties, of many, and here present them for general consideration in this treatise.

Were not the work indeed presented through you, my learned friends, I should scarce hope that it could come out scatheless and complete; for you have in general been the faithful witnesses of almost all the instances from which I have either collected the truth or confuted error. You have seen my dissections, and at my demonstrations of all that I maintain to be objects of sense, you have been accustomed to stand by and bear me out with your testimony. And as this book alone declares the blood to course and revolve by a new route, very different from the ancient and beaten pathway trodden for so many ages, and illustrated by such a host of learned and distinguished men, I was greatly afraid lest I might be charged with presumption did I lay my work before the public at home, or send it beyond seas for impression, unless I had first proposed the subject to you, had confirmed its conclusions by ocular demonstrations in your presence, had replied to your doubts and objections, and secured the assent and support of our distinguished President. For I was most intimately persuaded, that if I could make good my proposition before you and our College, illustrious by its numerous body of learned individuals, I had less to fear from others. I even ventured to hope that I should have the comfort of finding all that you had granted me in your sheer love of truth, conceded by others who were philosophers like yourselves. True philosophers, who are only eager for truth and knowledge, never regard themselves as already so thoroughly informed, but that they welcome further information from whomsoever and from wheresoever it may come; nor are they so narrow-minded as to imagine any of the arts or sciences transmitted to us by the ancients, in such a state of forwardness or completeness, that nothing is left for the ingenuity and industry of others. On the contrary, very many maintain that all we know is still infinitely less than all that still remains unknown; nor do philosophers pin their faith to others' precepts in such wise that they lose their liberty, and cease to give credence to the conclusions of their proper senses. Neither do they swear such fealty to their mistress Antiquity, that they openly, and in sight of all, deny and desert their friend Truth. But even as they see that the credulous and vain are disposed at the first blush to accept and believe everything that is proposed to them, so do they observe that the dull and unintellectual are indisposed to see what lies before their eyes, and even deny the light of the noonday sun. They teach us in our course of philosophy to sedulously avoid the fables of the poets and the fancies of the vulgar, as the false conclusions of the sceptics. And then the studious and good and true, never suffer their minds to be warped by the passions of hatred and envy, which unfit men duly to weigh the arguments that are advanced in behalf of truth, or to appreciate the proposition that is even fairly demonstrated. Neither do they think it unworthy of them to change their opinion if truth and undoubted demonstration require them to do so. They do not esteem it discreditable to desert error, though sanctioned by the highest antiquity, for they know full well that to err, to be deceived, is human; that many things are discovered by accident and that many may be learned indifferently from any quarter, by an old man from a youth, by a person of understanding from one of inferior capacity.

My dear colleagues, I had no purpose to swell this treatise into a large volume by quoting the names and writings of anatomists, or to make a parade of the strength of my memory, the extent of my reading, and the amount of my pains; because I profess both to learn and to teach anatomy, not from books but from dissections; not from the positions of philosophers but from the fabric of nature; and then because I do not think it right or proper to strive to take from the ancients any honor that is their due, nor yet to dispute with the moderns, and enter into controversy with those who have excelled in anatomy and been my teachers. I would not charge with wilful falsehood any one who was sincerely anxious for truth, nor lay it to any one's door as a crime that he had fallen into error. I avow myself the partisan of truth alone; and I can indeed say that I have used all my endeavours, bestowed all my pains on an attempt to produce something that should be agreeable to the good, profitable to the learned, and useful to letters.

Farewell, most worthy Doctors,

And think kindly of your Anatomist,

William Harvey.

INTRODUCTION

As we are about to discuss the motion, action, and use of the heart and arteries, it is imperative on us first to state what has been thought of these things by others in their writings, and what has been held by the vulgar and by tradition, in order that what is true may be confirmed, and what is false set right by dissection, multiplied experience, and accurate observation.

Almost all anatomists, physicians, and philosophers up to the present time have supposed, with Galen, that the object of the pulse was the same as that of respiration, and only differed in one particular, this being conceived to depend on the animal, the respiration on the vital faculty; the two, in all other respects, whether with reference to purpose or to motion, comporting themselves alike. Whence it is affirmed, as by Hieronymus Fabricius of Aquapendente, in his book on "Respiration," which has lately appeared, that as the pulsation of the heart and arteries does not suffice for the ventilation and refrigeration of the blood, therefore were the lungs fashioned to surround the heart. From this it appears that whatever has hitherto been said upon the systole and diastole, or on the motion of the heart and arteries, has been said with especial reference to the lungs.

But as the structure and movements of the heart differ from those of the lungs, and the motions of the arteries from those of the chest, so it seems likely that other ends and offices will thence arise, and that the pulsations and uses of the heart, likewise of the arteries, will differ in many respects from the heavings and uses of the chest and lungs. For did the arterial pulse and the respiration serve the same ends; did the arteries in their diastole take air into their cavities, as commonly stated, and in their systole emit fuliginous vapours by the same pores of the flesh and skin; and further, did they, in the time intermediate between the diastole and the systole, contain air, and at all times either air or spirits, or fuliginous vapours, what should then be said to Galen, who wrote a book on purpose to show that by nature the arteries contained blood, and nothing but blood, and consequently neither spirits nor air, as may readily be gathered from the experiments and reasonings contained in the same book? Now, if the arteries are filled in the diastole with air then taken into them (a larger quantity of air penetrating when the pulse is large and full), it must come to pass that if you plunge into a bath of water or of oil when the pulse is strong and full, it ought forthwith to become either smaller or much slower, since the circumambient bath will render it either difficult or impossible for the air to penetrate. In like manner, as all the arteries, those that are deep-seated as well as those that are superficial, are dilated at the same instant and with the same rapidity, how is it possible that air should penetrate to the deeper parts as freely and quickly through the skin, flesh, and other structures, as through the cuticle alone? And how should the arteries of the fœtus draw air into their cavities through the abdomen of the mother and the body of the womb? And how should seals, whales, dolphins, and other cetaceans, and fishes of every description, living in the depths of the sea, take in and emit air by the diastole and systole of their arteries through the infinite mass of water? For to say that they absorb the air that is present in the water, and emit their fumes into this medium, were to utter something like a figment. And if the arteries in their systole expel fuliginous vapours from their cavities through the pores of the flesh and skin, why not the spirits, which are said to be contained in those vessels, at the same time, since spirits are much more subtile than fuliginous vapours or smoke? And if the arteries take in and cast out air in the systole and diastole, like the lungs in the process of respiration, why do they not do the same thing when a wound is made in one of them, as in the operation of arteriotomy? When the windpipe is divided, it is sufficiently obvious that the air enters and returns through the wound by two opposite movements; but when an artery is divided, it is equally manifest that blood escapes in one continuous stream, and that no air either enters or issues. If the pulsations of the arteries fan and refrigerate the several parts of the body as the lungs do the heart, how comes it, as is commonly said, that the arteries carry the vital blood into the different parts, abundantly charged with vital spirits, which cherish the heat of these parts, sustain them when asleep, and recruit them when exhausted? How should it happen that, if you tie the arteries, immediately the parts not only become torpid, and frigid, and look pale, but at length cease even to be nourished? This, according to Galen, is because they are deprived of the heat which flowed through all parts from the heart, as its source; whence it would appear that the arteries rather carry warmth to the parts than serve for any fanning or refrigeration. Besides, how can their diastole draw spirits from the heart to warm the body and its parts, and means of cooling them from without? Still further, although some affirm that the lungs, arteries, and heart have all the same offices, they yet maintain that the heart is the workshop of the spirits, and that the arteries contain and transmit them; denying, however, in opposition to the opinion of Columbus, that the lungs can either make or contain spirits. They then assert, with Galen, against Erasistratus, that it is the blood, not spirits, which is contained in the arteries.

These opinions are seen to be so incongruous and mutually subversive, that every one of them is justly brought under suspicion. That it is blood and blood alone which is contained in the arteries is made manifest by the experiment of Galen, by arteriotomy, and by wounds; for from a single divided artery, as Galen himself affirms in more than one place, the whole of the blood may be withdrawn in the course of half an hour or less. The experiment of Galen alluded to is this: "If you include a portion of an artery between two ligatures, and slit it open lengthwise you will find nothing but blood"; and thus he proves that the arteries contain only blood. And we too may be permitted to proceed by a like train of reasoning: if we find the same blood in the arteries as in the veins, after having tied them in the same way, as I have myself repeatedly ascertained, both in the dead body and in living animals, we may fairly conclude that the arteries contain the same blood as the veins, and nothing but the same blood. Some, whilst they attempt to lessen the difficulty, affirm that the blood is spirituous and arterious, and virtually concede that the office of the arteries is to carry blood from the heart into the whole of the body, and that they are therefore filled with blood; for spirituous blood is not the less blood on that account. And no one denies the blood as such, even the portion of it which flows in the veins, is imbued with spirits. But if that portion of it which is contained in the arteries be richer in spirits, it is still to be believed that these spirits are inseparable from the blood, like those in the veins; that the blood and spirits constitute one body (like whey and butter in milk, or heat in hot water), with which the arteries are charged, and for the distribution of which from the heart they are provided. This body is nothing else than blood. But if this blood be said to be drawn from the heart into the arteries by the diastole of these vessels, it is then assumed that the arteries by their distension are filled with blood, and not with the surrounding air, as heretofore; for if they be said also to become filled with air from the ambient atmosphere, how and when, I ask, can they receive blood from the heart? If it be answered: during the systole, I take it to be impossible: the arteries would then have to fill while they contracted, to fill, and yet not become distended. But if it be said: during diastole, they would then, and for two opposite purposes, be receiving both blood and air, and heat and cold, which is improbable. Further when it is affirmed that the diastole of the heart and arteries is simultaneous, and the systole of the two is also concurrent, there is another incongruity. For how can two bodies mutually connected, which are simultaneously distended, attract or draw anything from one another? or being simultaneously contracted, receive anything from each other? And then it seems impossible that one body can thus attract another body into itself, so as to become distended, seeing that to be distended is to be passive, unless, in the manner of a sponge, which has been previously compressed by an external force, it is returning to its natural state. But it is difficult to conceive that there can be anything of this kind in the arteries. The arteries dilate, because they are filled like bladders or leathern bottles; they are not filled because they expand like bellows. This I think easy of demonstration, and indeed conceive that I have already proved it. Nevertheless, in that book of Galen headed "Quod Sanguis continetur in Arteriis," he quotes an experiment to prove the contrary. An artery having been exposed, is opened longitudinally, and a reed or other pervious tube is inserted into the vessel through the opening, by which the blood is prevented from being lost, and the wound is closed. "So long," he says, "as things are thus arranged, the whole artery will pulsate; but if you now throw a ligature about the vessel and tightly compress its wall over the tube, you will no longer see the artery beating beyond the ligature." I have never performed this experiment of Galen's nor do I think that it could very well be performed in the living body, on account of the profuse flow of blood that would take place from the vessel that was operated on; neither would the tube effectually close the wound in the vessel without a ligature; and I cannot doubt but that the blood would be found to flow out between the tube and the vessel. Still Galen appears by this experiment to prove both that the pulsative property extends from the heart by the walls of the arteries, and that the arteries, whilst they dilate, are filled by that pulsific force, because they expand like bellows, and do not dilate as if they are filled like skins, But the contrary is obvious in arteriotomy and in wounds; for the blood spurting from the arteries escapes with force, now farther, now not so far, alternately, or in jets; and the jet always takes place with the diastole of the artery, never with the systole. By which it clearly appears that the artery is dilated with the impulse of the blood; for of itself it would not throw the blood to such a distance and whilst it was dilating; it ought rather to draw air into its cavity through the wound, were those things true that are commonly stated concerning the uses of the arteries. Do not let the thickness of the arterial tunics impose upon us, and lead us to conclude that the pulsative property proceeds along them from the heart. For in several animals the arteries do not apparently differ from the veins; and in extreme parts of the body where the arteries are minutely subdivided, as in the brain, the hand, etc., no one could distinguish the arteries from the veins by the dissimilar characters of their coats: the tunics of both are identical. And then, in the aneurism proceeding from a wounded or eroded artery, the pulsation is precisely the same as in the other arteries, and yet it has no proper arterial covering. To this the learned Riolanus testifies along with me, in his Seventh Book.

Nor let any one imagine that the uses of the pulse and the respiration are the same, because, under the influences of the same causes, such as running, anger, the warm bath, or any other heating thing, as Galen says, they become more frequent and forcible together. For not only is experience in opposition to this idea, though Galen endeavours to explain it away, when we see that with excessive repletion the pulse beats more forcibly, whilst the respiration is diminished in amount· but in young persons the pulse is quick, whilst respiration is slow. So it is also in alarm, and amidst care, and under anxiety of mind; sometimes, too, in fevers, the pulse is rapid, but the respiration is slower than usual.

These and other objections of the same kind may be urged against the opinions mentioned. Nor are the views that are entertained of the offices and pulse of the heart, perhaps, less bound up with great and most inextricable difficulties. The heart, it is vulgarly said, is the fountain and workshop of the vital spirits, the centre from which life is dispensed to the several parts of the body. Yet it is denied that the right ventricle makes spirits, which is rather held to supply nourishment to the lungs. For these reasons it is maintained that fishes are without any right ventricle (and indeed every animal wants a right ventricle which is unfurnished with lungs), and that the right ventricle is present solely for the sake of the lungs.

1. Why, I ask, when we see that the structure of both ventricles is almost identical, there being the same apparatus of fibres, and braces, and valves, and vessels, and auricles, and both in the same way in our dissections are found to be filled up with blood similarly black in colour, and coagulated—why, I say, should their uses be imagined to be different, when the action, motion, and pulse of both are the same? If the three tricuspid valves placed at the entrance into the right ventricle prove obstacles to the reflux of the blood into the vena cava, and if the three semilunar valves which are situated at the commencement of the pulmonary artery be there, that they may prevent the return of the blood into the ventricle; why, when we find similar structures in connexion with the left ventricle, should we deny that they are there for the same end, of preventing here the egress, there the regurgitation, of the blood?

2. And, when we have these structures, in points of size, form, and situation, almost in every respect the same in the left as in the right ventricle, why should it be said that things are arranged in the former for the egress and regress of spirits,

and in the latter or right ventricle, for the blood? The same arrangement cannot be held fitted to favour or impede the motion of the blood and of spirits indifferently.

3. And when we observe that the passages and vessels are severally in relation to one another in point of size, viz., the pulmonary artery to the pulmonary veins; why should the one be destined to a private purpose, that of furnishing the lungs, the other to a public function?

4. And as Realdus Columbus says, is it probable that such a quantity of blood should be required for the nutrition of the lungs; the vessel that leads to them, the vena arteriosa or pulmonary artery being of greater capacity than both the iliac veins?

5. And I ask, as the lungs are so close at hand, and in continual motion, and the vessel that supplies them is of such dimensions, what is the use or meaning of this pulse of the right ventricle? and why was nature reduced to the necessity of adding another ventricle for the sole purpose of nourishing the lungs?

When it is said that the left ventricle draws materials for the formation of spirits, air and blood, from the lungs and right sinuses of the heart, and in like manner sends spirituous blood into the aorta, drawing fuliginous vapours from there, and sending them by the pulmonary vein into the lungs, whence spirits are at the same time obtained for transmission into the aorta, I ask how, and by what means is the separation effected? And how comes it that spirits and fuliginous vapours can pass hither and thither without admixture or confusion? If the mitral cuspidate valves do not prevent the egress of fuliginous vapours to the lungs, how should they oppose the escape of air? And how should the semilunars hinder the regress of spirits from the aorta upon each supervening diastole of the heart? Above all, how can they say that the spirituous blood is sent from the pulmonary veins by the left ventricle into the lungs without any obstacle to its passage from the mitral valves, when they have previously asserted that the air entered by the same vessel from the lungs into the left ventricle, and have brought forward these same mitral valves as obstacles to its retrogression? Good God! how should the mitral valves prevent the regurgitation of air and not of blood?

Moreover, when they appoint the pulmonary artery, a vessel of great size, with the coverings of an artery, to none but a kind of private and single purpose, that, namely, of nourishing the lungs, why should the pulmonary vein, which is scarcely so large, which has the coats of a vein, and is soft and lax, be presumed to be made for many—three or four different—uses? For they will have it that air passes through this vessel from the lungs into the left ventricle; that fuliginous vapours escape by it from the heart into the lungs; and that a portion of the spirituous blood is distributed to the lungs for their refreshment.

If they will have it that fumes and air—fumes flowing from, air proceeding towards the heart—are transmitted by the same conduit, I reply, that nature is not wont to construct but one vessel, to contrive but one way for such contrary motions and purposes, nor is anything of the kind seen elsewhere.

If fumes or fuliginous vapours and air permeate this vessel, as they do the pulmonary bronchia, wherefore do we find neither air nor fuliginous vapours when we divide the pulmonary vein? Why do we always find this vessel full of sluggish blood, never of air, whilst in the lungs we find abundance of air remaining?

If any one will perform Galen's experiment of dividing the trachea of a living dog, forcibly distending the lungs with a pair of bellows, and then tying the trachea securely, he will find, when he has laid open the thorax, abundance of air in the lungs, even to their extreme investing tunic, but none in either the pulmonary veins or the left ventricle of the heart. But did the heart either attract air from the lungs, or did the lungs transmit any air to the heart, in the living dog, much more ought this to be the case in the experiment just referred to. Who, indeed, doubts that, did he inflate the lungs of a subject in the dissecting-room, he would instantly see the air making its way by this route, were there actually any such passage for it? But this office of the pulmonary veins, namely, the transference of air from the lungs of the heart, is held of such importance, that Hieronymus Fabricius of Aquapendente, contends that the lungs were made for the sake of this vessel, and that it constitutes the principal element in their structure.

But I should like to be informed why, if the pulmonary vein were destined for the conveyance of air, it has the structure of a blood-vessel here. Nature had rather need of annular tubes, such as those of the bronchi, in order that they might always remain open, and not be liable to collapse; and that they might continue entirely free from blood, lest the liquid should interfere with the passage of the air, as it so obviously does when the lungs labour from being either greatly oppressed or loaded in a less degree with phlegm, as they are when the breathing is performed with a sibilous or rattling noise.

Still less is that opinion to be tolerated which, as a two-fold material, one aerial, one sanguineous, is required for the composition of vital spirits, supposes the blood to ooze through the septum of the heart from the right to the left ventricle by certain hidden porosities, and the air to be attracted from the lungs through the great vessel, the pulmonary vein; and which, consequently, will have it, that there are numerous porosities in the septum of the heart adapted for the transmission of the blood. But by Hercules! no such pores can be demonstrated, nor in fact do any such exist. For the septum of the heart is of a denser and more compact structure than any portion of the body, except the bones and sinews. But even supposing that there were foramina or pores in this situation, how could one of the ventricles extract anything from the other—the left, *e.g.*, obtain blood from the right, when we see that both ventricles contract and dilate simultaneously? Why should we not rather believe that the right took spirits from the left, than that the left obtained blood from the right ventricle through these foramina? But it is certainly mysterious and incongruous that blood should be supposed to be most commodiously drawn through a set of obscure or invisible ducts, and air through perfectly open passages, at one and the same moment. And why, I ask, is recourse had to secret and invisible porosities, to uncertain and obscure channels, to explain the passage of the blood into the left ventricle, when there is so open a way through the pulmonary veins? I own it has always appeared extraordinary to me that they should have chosen to make, or rather to imagine, a way through the thick, hard, dense, and most compact septum of the heart, rather than take that by the open pulmonary vein, or even through the lax, soft and spongy substance of the lungs at large. Besides, if the blood could permeate the substance of the septum, or could be imbibed from the ventricles, what use were there for the coronary artery and vein, branches of which proceed to the septum itself, to supply it with nourishment? And what is especially worthy of notice is this: if in the fœtus, where everything is more lax and soft, nature saw herself reduced to the necessity of bringing the blood from the right to the left side of the heart by the foramen ovale, from the vena cava through the pulmonary vein, how should it be likely that in the adult she should pass it so commodiously, and without an effort through the septum of the ventricles which has now become denser by age?

Andreas Laurentius,[1] resting on the authority of Galen[2] and the experience of Hollerius, asserts and proves that the serum and pus in empyema, absorbed from the cavities of the chest into the pulmonary vein may be expelled and got rid of with the urine and fæces through the left ventricle of the heart and arteries. He quotes the case of a certain person affected with melancholia, and who suffered from repeated fainting fits, who was relieved from the paroxysms on passing a quantity of turbid, fetid and acrid urine. But he died at last, worn out by disease; and when the body came to be opened after death, no fluid like that he had micturated was discovered either in the bladder or the kidneys; but in the left ventricle of the heart and cavity of the thorax plenty of it was met with. And then Laurentius boasts that he had predicted the cause of the symptoms. For my own part, however, I cannot but wonder, since he had divined and predicted that heterogeneous matter could be discharged by the course he indicates, why he could not or would not perceive, and inform us that, in the natural state of things, the blood might be commodiously transferred from the lungs to the left ventricle of the heart by the very same route.

Since, therefore, from the foregoing considerations and many others to the same effect, it is plain that what has heretofore been said concerning the motion and function of the heart and arteries must appear obscure, inconsistent, or even impossible to him who carefully considers the entire subject, it would be proper to look more narrowly into the matter to contemplate the motion of the heart and arteries, not only in man, but in all animals that have hearts; and also, by frequent appeals to vivisection, and much ocular inspection, to investigate and discern the truth.

[1] Lib. ix, cap. xi, quest. 12.

[2] De Locis Affectis. lib. vi, cap. 7.

ON THE MOTION OF THE HEART AND BLOOD IN ANIMALS

CHAPTER I

The Author's Motives for Writing

WHEN I first gave my mind to vivisections, as a means of discovering the motions and uses of the heart, and sought to discover these from actual inspection, and not from the writings of others, I found the task so truly arduous, so full of difficulties, that I was almost tempted to think, with Fracastorius, that the motion of the heart was only to be comprehended by God. For I could neither rightly perceive at first when the systole and when the diastole took place, nor when and where dilatation and contraction occurred, by reason of the rapidity of the motion, which in many animals is accomplished in the twinkling of an eye, coming and going like a flash of lightning; so that the systole presented itself to me now from this point, now from that; the diastole the same; and then everything was reversed, the motions occurring, as it seemed, variously and confusedly together. My mind was therefore greatly unsettled nor did I know what I should myself conclude, nor what believe from others. I was not surprised that Andreas Laurentius should have written that the motion of the heart was as perplexing as the flux and reflux of Euripus had appeared to Aristotle.

At length, by using greater and daily diligence and investigation, making frequent inspection of many and various animals, and collating numerous observations, I thought that I had attained to the truth, that I should extricate myself and escape from this labyrinth, and that I had discovered what I so much desired, both the motion and the use of the heart and arteries. From that time I have not hesitated to expose my views upon these subjects, not only in private to my friends, but also in public, in my anatomical lectures, after the manner of the Academy of old.

These views as usual, pleased some more, others less; some chid and calumniated me, and laid it to me as a crime that I had dared to depart from the precepts and opinions of all anatomists; others desired further explanations of the novelties, which they said were both worthy of consideration, and might perchance be found of signal use. At length, yielding to the requests of my friends, that all might be made participators in my labors, and partly moved by the envy of others, who, receiving my views with uncandid minds and understanding them indifferently, have essayed to traduce me publicly, I have moved to commit these things to the press, in order that all may be enabled to form an opinion both of me and my labours. This step I take all the more willingly, seeing that Hieronymus Fabricius of Aquapendente, although he has accurately and learnedly delineated almost every one of the several parts of animals in a special work, has left the heart alone untouched. Finally, if any use or benefit to this department of the republic of letters should accrue from my labours, it will, perhaps, be allowed that I have not lived idly, and as the old man in the comedy says:

> For never yet hath any one attained
> To such perfection, but that time, and place,
> And use, have brought addition to his knowledge;
> Or made correction, or admonished him,
> That he was ignorant of much which he
> Had thought he knew; or led him to reject
> What he had once esteemed of highest price.

So will it, perchance, be found with reference to the heart at this time; or others, at least, starting hence, with the way pointed out to them, advancing under the guidance of a happier genius, may make occasion to proceed more fortunately, and to inquire more accurately.

CHAPTER II

On the Motions of the Heart as Seen in the Dissection of Living Animals

In the first place, then, when the chest of a living animal is laid open and the capsule that immediately surrounds the heart is slit up or removed, the organ is seen now to move, now to be at rest; there is a time when it moves, and a time when it is motionless.

These things are more obvious in the colder animals, such as toads, frogs, serpents, small fishes, crabs, shrimps, snails, and shell-fish. They also become more distinct in warm-blooded animals, such as the dog and hog, if they be attentively noted when the heart begins to flag, to move more slowly, and, as it were, to die: the movements then become slower and rarer, the pauses longer, by which it is made much more easy to perceive and unravel what the motions really are, and how they are performed. In the pause, as in death, the heart is soft, flaccid, exhausted, lying, as it were, at rest.

In the motion, and interval in which this is accomplished, three principal circumstances are to be noted:

1. That the heart is erected, and rises upwards to a point, so that at this time it strikes against the breast and the pulse is felt externally.

2. That it is everywhere contracted, but more especially towards the sides so that it looks narrower, relatively longer, more drawn together. The heart of an eel taken out of the body of the animal and placed upon the table or the hand, shows these particulars; but the same things are manifest in the hearts of all small fishes and of those colder animals where the organ is more conical or elongated.

3. The heart being grasped in the hand, is felt to become harder during its action. Now this hardness proceeds from tension, precisely as when the forearm is grasped, its tendons are perceived to become tense and resilient when the fingers are moved.

4. It may further be observed in fishes, and the colder blooded animals, such as frogs, serpents, etc., that the heart,

when it moves, becomes of a paler color, when quiescent of a deeper blood-red color.

From these particulars it appears evident to me that the motion of the heart consists in a certain universal tension—both contraction in the line of its fibres, and constriction in every sense. It becomes erect, hard, and of diminished size during its action; the motion is plainly of the same nature as that of the muscles when they contract in the line of their sinews and fibres; for the muscles, when in action, acquire vigor and tenseness, and from soft become hard, prominent, and thickened: and in the same manner the heart.

We are therefore authorized to conclude that the heart, at the moment of its action, is at once constricted on all sides, rendered thicker in its parietes and smaller in its ventricles, and so made apt to project or expel its charge of blood. This, indeed, is made sufficiently manifest by the preceding fourth observation in which we have seen that the heart, by squeezing out the blood that it contains, becomes paler, and then when it sinks into repose and the ventricle is filled anew with blood, that the deeper crimson colour returns. But no one need remain in doubt of the fact, for if the ventricle be pierced the blood will be seen to be forcibly projected outwards upon each motion or pulsation when the heart is tense.

These things, therefore, happen together or at the same instant: the tension of the heart, the pulse of its apex, which is felt externally by its striking against the chest, the thickening of its parietes, and the forcible expulsion of the blood it contains by the constriction of its ventricles.

Hence the very opposite of the opinions commonly received appears to be true; inasmuch as it is generally believed that when the heart strikes the breast and the pulse is felt without, the heart is dilated in its ventricles and is filled with blood; but the contrary of this is the fact, and the heart, when it contracts (and the impulse of the apex is conveyed through the chest wall), is emptied. Whence the motion which is generally regarded as the diastole of the heart, is in truth its systole. And in like manner the intrinsic motion of the heart is not the diastole but the systole; neither is it in the diastole that the heart grows firm and tense, but in the systole, for then only, when tense, is it moved and made vigorous.

Neither is it by any means to be allowed that the heart only moves in the lines of its straight fibres, although the great Vesalius giving this notion countenance, quotes a bundle of osiers bound in a pyramidal heap in illustration; meaning, that as the apex is approached to the base, so are the sides made to bulge out in the fashion of arches, the cavities to dilate, the ventricles to acquire the form of a cupping-glass and so to suck in the blood. But the true effect of every one of its fibres is to constringe the heart at the same time they render it tense; and this rather with the effect of thickening and amplifying the walls and substance of the organ than enlarging its ventricles. And, again, as the fibres run from the apex to the base, and draw the apex towards the base, they do not tend to make the walls of the heart bulge out in circles, but rather the contrary; inasmuch as every fibre that is circularly disposed, tends to become straight when it contracts; and is distended laterally and thickened, as in the case of muscular fibres in general, when they contract, that is, when they are shortened longitudinally, as we see them in the bellies of the muscles of the body at large. To all this let it be added, that not only are the ventricles contracted in virtue of the direction and condensation of their walls, but farther, that those fibres, or bands, styled nerves by Aristotle, which are so conspicuous in the ventricles of the larger animals, and contain all the straight fibres (the parietes of the heart containing only circular ones), when they contract simultaneously by an admirable adjustment all the internal surfaces are drawn together as if with cords, and so is the charge of blood expelled with force.

Neither is it true, as vulgarly believed, that the heart by any dilatation or motion of its own, has the power of drawing the blood into the ventricles; for when it acts and becomes tense, the blood is expelled; when it relaxes and sinks together it receives the blood in the manner and wise which will by-and-by be explained.

CHAPTER III

Of the Motions of the Arteries, as Seen in the Dissection of Living Animals

In connexion with the motions of the heart these things are further to be observed having reference to the motions and pulses of the arteries.

1. At the moment the heart contracts, and when the breast is struck, when in short the organ is in its state of systole, the arteries are dilated, yield a pulse, and are in the state of diastole. In like manner, when the right ventricle contracts and propels its charge of blood, the pulmonary artery is distended at the same time with the other arteries of the body.

2. When the left ventricle ceases to act, to contract, to pulsate, the pulse in the arteries also ceases; further, when this ventricle contracts languidly, the pulse in the arteries is scarcely perceptible. In like manner, the pulse in the right ventricle failing, the pulse in the pulmonary artery ceases also.

3. Further, when an artery is divided or punctured, the blood is seen to be forcibly propelled from the wound the moment the left ventricle contracts; and, again, when the pulmonary artery is wounded, the blood will be seen spouting forth with violence at the instant when the right ventricle contracts.

So also in fishes, if the vessel which leads from the heart to the gills be divided, at the moment when the heart becomes tense and contracted, at the same moment does the blood flow with force from the divided vessel.

In the same way, when we see the blood in arteriotomy projected now to a greater, now to a less distance, and that the greater jet corresponds to the diastole of the artery and to the time when the heart contracts and strikes the ribs, and is in its state of systole, we understand that the blood is expelled by the same movement.

From these facts it is manifest, in opposition to commonly received opinions, that the diastole of the arteries corresponds with the time of the heart's systole; and that the arteries are filled and distended by the blood forced into them by the contraction of the ventricles; the arteries, therefore, are distended, because they are filled like sacs or bladders, and are not filled because they expand like bellows. It is in virtue of one and the same cause, therefore, that all the arteries of the body pulsate, viz., the contraction of the left ventricle; in the same way as the pulmonary artery pulsates by the contraction of the right ventricle.

Finally, that the pulses of the arteries are due to the impulses of the blood from the left ventricle, may be illustrated by blowing into a glove, when the whole of the fingers will be found to become distended at one and the same time, and in their tension to bear some resemblance to the pulse. For in the ratio of the tension is the pulse of the heart, fuller, stronger, and more frequent as that acts more vigorously, still preserving the rhythm and volume, and order of the heart's contractions. Nor is it to be expected that because of the motion of the blood, the time at which the contraction of the heart takes place, and that at which the pulse in an artery (especially a distant one) is felt, shall be otherwise than simultaneous: it is here the same as in blowing up a glove or bladder; for in a plenum (as in a drum, a long piece of timber, etc.) the stroke and the motion occur at both extremities at the same time. Aristotle,[1] too, has said, "the blood of all animals palpitates within their veins (meaning the arteries), and by the pulse is sent everywhere simultaneously." And further,[2] "thus do all the veins pulsate together and by successive strokes, because they all depend upon the heart; and, as it is always in motion, so are they likewise always moving together, but by successive movements." It is well to observe with Galen, in this place, that the old philosophers called the arteries veins.

I happened upon one occasion to have a particular case under my care, which plainly satisfied me of the truth: A certain person was affected with a large pulsating tumour on the right side of the neck, called an aneurism, just at that part where the artery descends into the axilla, produced by an erosion of the artery itself, and daily increasing in size;

[1] De Anim., iii, cap. 9. [2] De Respir., cap. 20.

this tumour was visibly distended as it received the charge of blood brought to it by the artery, with each stroke of the heart; the connexion of parts was obvious when the body of the patient came to be opened after his death. The pulse in the corresponding arm was small, in consequence of the greater portion of the blood being diverted into the tumour and so intercepted.

Whence it appears that whenever the motion of the blood through the arteries is impeded, whether it be by compression or infarction, or interception, there do the remote divisions of the arteries beat less forcibly, seeing that the pulse of the arteries is nothing more than the impulse or shock of the blood in these vessels.

CHAPTER IV

Of the Motion of the Heart and Its Auricles, as Seen in the Bodies of Living Animals

Besides the motions already spoken of, we have still to consider those that appertain to the auricles.

Caspar Bauhin and John Riolan,[1] most learned men and skilful anatomists, inform us that from their observations, that if we carefully watch the movements of the heart in the vivisection of an animal, we shall perceive four motions distinct in time and in place, two of which are proper to the auricles, two to the ventricles. With all deference to such authority I say that there are four motions distinct in point of place, but not of time; for the two auricles move together, and so also do the two ventricles, in such wise that though the places be four, the times are only two. And this occurs in the following manner:

There are, as it were, two motions going on together: one of the auricles, another of the ventricles; these by no means taking place simultaneously, but the motion of the auricles preceding, that of the heart following; the motion appearing to begin from the auricles and to extend to the ventricles. When all things are becoming languid, and the heart is dying, as also in fishes and the colder blooded animals there

[1] Bauhin, lib. ii, cap. 21. Riolan, lib. viii, cap. 1.

is a short pause between these two motions, so that the heart aroused, as it were, appears to respond to the motion, now more quickly, now more tardily; and at length, when near to death, it ceases to respond by its proper motion, but seems, as it were, to nod the head, and is so slightly moved that it appears rather to give signs of motion to the pulsating auricles than actually to move. The heart, therefore, ceases to pulsate sooner than the auricles, so that the auricles have been said to outlive it, the left ventricle ceasing to pulsate first of all; then its auricle, next the right ventricle; and, finally, all the other parts being at rest and dead, as Galen long since observed, the right auricle still continues to beat; life, therefore, appears to linger longest in the right auricle. Whilst the heart is gradually dying, it is sometimes seen to reply, after two or three contractions of the auricles, roused as it were to action, and making a single pulsation, slowly, unwillingly, and with an effort.

But this especially is to be noted, that after the heart has ceased to beat, the auricles however still contracting, a finger placed upon the ventricles perceives the several pulsations of the auricles, precisely in the same way and for the same reason, as we have said, that the pulses of the ventricles are felt in the arteries, to wit, the distension produced by the jet of blood. And if at this time, the auricles alone pulsating, the point of the heart be cut off with a pair of scissors, you will perceive the blood flowing out upon each contraction of the auricles. Whence it is manifest that the blood enters the ventricles, not by any attraction or dilatation of the heart, but by being thrown into them by the pulses of the auricles.

And here I would observe, that whenever I speak of pulsations as occurring in the auricles or ventricles, I mean contractions: first the auricles contract, and then and subsequently the heart itself contracts. When the auricles contract they are seen to become whiter, especially where they contain but little blood; but they are filled as magazines or reservoirs of the blood, which is tending spontaneously and, by its motion in the veins, under pressure towards the centre; the whiteness indicated is most conspicuous towards the extremities or edges of the auricles at the time of their contractions.

In fishes and frogs, and other animals which have hearts with but a single ventricle, and for an auricle have a kind of bladder much distended with blood, at the base of the organ, you may very plainly perceive this bladder contracting first, and the contraction of the heart or ventricle following afterwards.

But I think it right to describe what I have observed of an opposite character: the heart of an eel, of several fishes, and even of some (of the higher) animals taken out of the body, pulsates without auricles; nay, if it be cut in pieces the several parts may still be seen contracting and relaxing; so that in these creatures the body of the heart may be seen pulsating and palpitating, after the cessation of all motion in the auricle. But is not this perchance peculiar to animals more tenacious of life, whose radical moisture is more glutinous, or fat and sluggish, and less readily soluble? The same faculty indeed appears in the flesh of eels, which even when skinned and embowelled, and cut into pieces, are still seen to move.

Experimenting with a pigeon upon one occasion, after the heart had wholly ceased to pulsate, and the auricles too had become motionless, I kept my finger wetted with saliva and warm for a short time upon the heart, and observed that under the influence of this fomentation it recovered new strength and life, so that both ventricles and auricles pulsated, contracting and relaxing alternately, recalled as it were from death to life.

Besides this, however, I have occasionally observed, after the heart and even its right auricle had ceased pulsating,—when it was in articulo mortis in short,—that an obscure motion, an undulation or palpitation, remained in the blood itself, which was contained in the right auricle, this being apparent so long as it was imbued with heat and spirit. And, indeed, a circumstance of the same kind is extremely manifest in the course of the generation of animals, as may be seen in the course of the first seven days of the incubation of the chick: A drop of blood makes its appearance which palpitates, as Aristotle had already observed; from this, when the growth is further advanced and the chick is fashioned, the auricles of the heart are formed, which pulsating henceforth give constant signs of life. When at length, and after the lapse of a few days, the outline of the body begins to be distinguished, then is the ventricular part of the heart also produced, but it continues for a time white and apparently bloodless, like the rest of the animal; neither does it pulsate or give signs of motion. I have seen a similar condition of the heart in the human fœtus about the beginning of the third month, the heart then being whitish and bloodless, although its auricles contained a considerable quantity of purple blood. In the same way in the egg, when the chick was formed and had increased in size, the heart too increased and acquired ventricles, which then began to receive and to transmit blood.

And this leads me to remark that he who inquires very particularly into this matter will not conclude that the heart, as a whole, is the primum vivens, ultimum moriens,—the first part to live, the last to die,—but rather its auricles, or the part which corresponds to the auricles in serpents, fishes, etc., which both lives before the heart and dies after it.

Nay, has not the blood itself or spirit an obscure palpitation inherent in it, which it has even appeared to me to retain after death? and it seems very questionable whether or not we are to say that life begins with the palpitation or beating of the heart. The seminal fluid of all animals—the prolific spirit, as Aristotle observed, leaves their body with a bound and like a living thing; and nature in death, as Aristotle[2] further remarks, retracing her steps, reverts to where she had set out, and returns at the end of her course to the goal whence she had started. As animal generation proceeds from that which is not animal, entity from nonentity, so, by a retrograde course, entity, by corruption, is resolved into nonentity, whence that in animals, which was last created, fails first and that which was first, fails last.

I have also observed that almost all animals have truly a heart, not the larger creatures only, and those that have red blood, but the smaller, and pale-blooded ones also, such as slugs, snails, scallops, shrimps, crabs, crayfish, and many others; nay, even in wasps, hornets, and flies, I have, with

[2] De Motu Animal., cap. 8.

the aid of a magnifying glass, and at the upper part of what is called the tail, both seen the heart pulsating myself, and shown it to many others.

But in the pale-blooded tribes the heart pulsates sluggishly and deliberately, contracting slowly as in animals that are moribund, a fact that may readily be seen in the snail, whose heart will be found at the bottom of that orifice in the right side of the body which is seen to be opened and shut in the course of respiration, and whence saliva is discharged, the incision being made in the upper aspect of the body, near the part which corresponds to the liver.

This, however, is to be observed: that in winter and the colder season, exsanguine animals, such as the snail, show no pulsation; they seem rather to live after the manner of vegetables, or of those other productions which are therefore designated plant-animals.

It is also to be noted that all animals which have a heart have also auricles, or something analogous to auricles; and, further, that whenever the heart has a double ventricle, there are always two auricles present, but not otherwise. If you turn to the production of the chick in ovo, however, you will find at first no more a vesicle or auricle, or pulsating drop of blood; it is only by and by, when the development has made some progress, that the heart is fashioned; even so in certain animals not destined to attain to the highest perfection in their organization, such as bees, wasps, snails, shrimps, crayfish, etc., we only find a certain pulsating vesicle, like a sort of red or white palpitating point, as the beginning or principle of their life.

We have a small shrimp in these countries, which is taken in the Thames and in the sea, the whole of whose body is transparent; this creature, placed in a little water, has frequently afforded myself and particular friends an opportunity of observing the motions of the heart with the greatest distinctness, the external parts of the body presenting no obstacle to our view, but the heart being perceived as though it had been seen through a window.

I have also observed the first rudiments of the chick in the course of the fourth or fifth day of the incubation, in the guise of a little cloud, the shell having been removed and the egg immersed in clear tepid water. In the midst of the cloudlet in question there was a bloody point so small that it disappeared during the contraction and escaped the sight, but in the relaxation it reappeared again, red and like the point of a pin; so that betwixt the visible and invisible, betwixt being and not being, as it were, it gave by its pulses a kind of representation of the commencement of life.

CHAPTER V

Of the Motion, Action and Office of the Heart

From these and other observations of a similar nature, I am persuaded it will be found that the motion of the heart is as follows:

First of all, the auricle contracts, and in the course of its contraction forces the blood (which it contains in ample quantity as the head of the veins, the store-house and cistern of the blood) into the ventricle, which, being filled, the heart raises itself straightway, makes all its fibres tense, contracts the ventricles, and performs a beat, by which beat it immediately sends the blood supplied to it by the auricle into the arteries. The right ventricle sends its charge into the lungs by the vessel which is called vena arteriosa, but which in structure and function, and all other respects, is an artery. The left ventricle sends its charge into the aorta, and through this by the arteries to the body at large.

These two motions, one of the ventricles, the other of the auricles, take place consecutively, but in such a manner that there is a kind of harmony or rhythm preserved between them, the two concurring in such wise that but one motion is apparent, especially in the warmer blooded animals, in which the movements in question are rapid. Nor is this for any other reason than it is in a piece of machinery, in which, though one wheel gives motion to another, yet all the wheels seem to move simultaneously; or in that mechanical contrivance which is adapted to firearms, where, the trigger being touched, down comes the flint, strikes against the steel, elicits a spark, which falling among the powder, ignites it, when the flame extends, enters the barrel, causes the explosion, propels the ball, and the mark is attained—all of which incidents, by reason of the celerity with which they happen, seem to take place in the twinkling of an eye. So also in deglutition: by the elevation of the root of the tongue, and the compression of the mouth, the food or drink is pushed into the fauces, when the larynx is closed by its muscles and by the epiglottis. The pharynx is then raised and opened by its muscles in the same way as a sac that is to be filled is lifted up and its mouth dilated. Upon the mouthful being received, it is forced downwards by the transverse muscles, and then carried farther by the longitudinal ones. Yet all these motions, though executed by different and distinct organs, are performed harmoniously, and in such order that they seem to constitute but a single motion and act, which we call deglutition.

Even so does it come to pass with the motions and action of the heart, which constitute a kind of deglutition, a transfusion of the blood from the veins to the arteries. And if anyone, bearing these things in mind, will carefully watch the motions of the heart in the body of a living animal, he will perceive not only all the particulars I have mentioned, viz., the heart becoming erect, and making one continuous motion with its auricles; but farther, a certain obscure undulation and lateral inclination in the direction of the axis of the right ventricle, as if twisting itself slightly in performing its work. And indeed everyone may see, when a horse drinks, that the water is drawn in and transmitted to the stomach at each movement of the throat, which movement produces a sound and yields a pulse both to the ear and the touch; in the same way it is with each motion of the heart, when there is the delivery of a quantity of blood from the veins to the arteries a pulse takes place, and can be heard within the chest.

The motion of the heart, then, is entirely of this description, and the one action of the heart is the transmission of the blood and its distribution, by means of the arteries, to the very extremities of the body; so that the pulse which we feel in the arteries is nothing more than the impulse of the blood derived from the heart.

Whether or not the heart, besides propelling the blood,

giving it motion locally, and distributing it to the body, adds anything else to it—heat, spirit, perfection,—must be inquired into by-and-by, and decided upon other grounds. So much may suffice at this time, when it is shown that by the action of the heart the blood is transfused through the ventricles from the veins to the arteries, and distributed by them to all parts of the body.

The above, indeed, is admitted by all, both from the structure of the heart and the arrangement and action of its valves. But still they are like persons purblind or groping about in the dark, for they give utterance to various, contradictory, and incoherent sentiments, delivering many things upon conjecture, as we have already shown.

The grand cause of doubt and error in this subject appears to me to have been the intimate connexion between the heart and the lungs. When men saw both the pulmonary artery and the pulmonary veins losing themselves in the lungs, of course it became a puzzle to them to know how or by what means the right ventricle should distribute the blood to the body, or the left draw it from the venæ cavæ. This fact is borne witness to by Galen, whose words, when writing against Erasistratus in regard to the origin and use of the veins and the coction of the blood, are the following[1]: "You will reply," he says, "that the effect is so; that the blood is prepared in the liver, and is thence transferred to the heart to receive its proper form and last perfection; a statement which does not appear devoid of reason; for no great and perfect work is ever accomplished at a single effort, or receives its final polish from one instrument. But if this be actually so, then show us another vessel which draws the absolutely perfect blood from the heart, and distributes it as the arteries do the spirits over the whole body." Here then is a reasonable opinion not allowed, because, forsooth, besides not seeing the true means of transit, he could not discover the vessel which should transmit the blood from the heart to the body at large!

But had anyone been there in behalf of Erasistratus, and of that opinion which we now espouse, and which Galen himself acknowledges in other respects consonant with reason, to

[1] De Placitis Hippocratis et Platonis, vi.

have pointed to the aorta as the vessel which distributes the blood from the heart to the rest of the body, I wonder what would have been the answer of that most ingenious and learned man? Had he said that the artery transmits spirits and not blood, he would indeed sufficiently have answered Erasistratus, who imagined that the arteries contained nothing but spirits; but then he would have contradicted himself, and given a foul denial to that for which he had keenly contended in his writings against this very Erasistratus, to wit, that blood in substance is contained in the arteries, and not spirits; a fact which he demonstrated not only by many powerful arguments, but by experiments.

But if the divine Galen will here allow, as in other places he does, "that all the arteries of the body arise from the great artery, and that this takes its origin from the heart; that all these vessels naturally contain and carry blood; that the three semilunar valves situated at the orifice of the aorta prevent the return of the blood into the heart, and that nature never connected them with this, the most noble viscus of the body, unless for some important end"; if, I say, this father of physicians concedes all these things,—and I quote his own words,—I do not see how he can deny that the great artery is the very vessel to carry the blood, when it has attained its highest term of perfection, from the heart for distribution to all parts of the body. Or would he perchance still hesitate, like all who have come after him, even to the present hour, because he did not perceive the route by which the blood was transferred from the veins to the arteries, in consequence, as I have already said, of the intimate connexion between the heart and the lungs? And that this difficulty puzzled anatomists not a little, when in their dissections they found the pulmonary artery and left ventricle full of thick, black, and clotted blood, plainly appears, when they felt themselves compelled to affirm that the blood made its way from the right to the left ventricle by transuding through the septum of the heart. But this fancy I have already refuted. A new pathway for the blood must therefore be prepared and thrown open, and being once exposed, no further difficulty will, I believe, be experienced by anyone in admitting what I have already proposed in regard to the pulse of the heart and arteries, viz., the passage of the blood from the veins to the arteries, and its distribution to the whole of the body by means of these vessels.

CHAPTER VI

Of the Course by which the Blood is Carried from the Vena Cava into the Arteries, or from the Right into the Left Ventricle of the Heart

Since the intimate connexion of the heart with the lungs, which is apparent in the human subject, has been the probable cause of the errors that have been committed on this point, they plainly do amiss who, pretending to speak of the parts of animals generally, as anatomists for the most part do, confine their researches to the human body alone, and that when it is dead. They obviously do not act otherwise than he who, having studied the forms of a single commonwealth, should set about the composition of a general system of polity; or who, having taken cognizance of the nature of a single field, should imagine that he had mastered the science of agriculture; or who, upon the ground of one particular proposition, should proceed to draw general conclusions.

Had anatomists only been as conversant with the dissection of the lower animals as they are with that of the human body, the matters that have hitherto kept them in a perplexity of doubt would, in my opinion, have met them freed from every kind of difficulty.

And first, in fishes, in which the heart consists of but a single ventricle, being devoid of lungs, the thing is sufficiently manifest. Here the sac, which is situated at the base of the heart, and is the part analogous to the auricle in man, plainly forces the blood into the heart, and the heart, in its turn, conspicuously transmits it by a pipe or artery, or vessel analogous to an artery; these are facts which are confirmed by simple ocular inspection, as well as by a division of the vessel, when the blood is seen to be projected by each pulsation of the heart.

The same thing is also not difficult of demonstration in

those animals that have, as it were, no more than a single ventricle to the heart, such as toads, frogs, serpents, and lizards, which have lungs in a certain sense, as they have a voice. I have many observations by me on the admirable structure of the lungs of these animals, and matters appertaining, which, however, I cannot introduce in this place. Their anatomy plainly shows us that the blood is transferred in them from the veins to the arteries in the same manner as in higher animals, viz., by the action of the heart; the way, in fact, is patent, open, manifest; there is no difficulty, no room for doubt about it; for in them the matter stands precisely as it would in man were the septum of his heart perforated or removed, or one ventricle made out of two; and this being the case, I imagine that no one will doubt as to the way by which the blood may pass from the veins into the arteries.

But as there are actually more animals which have no lungs than there are furnished with them, and in like manner a greater number which have only one ventricle than there are with two, it is open to us to conclude, judging from the mass or multitude of living creatures, that for the major part, and generally, there is an open way by which the blood is transmitted from the veins through the sinuses or cavities of the heart into the arteries.

I have, however, cogitating with myself, seen further, that the same thing obtained most obviously in the embryos of those animals that have lungs; for in the fœtus the four vessels belonging to the heart, viz., the vena cava, the pulmonary artery, the pulmonary vein, and the great artery or aorta, are all connected otherwise than in the adult, a fact sufficiently known to every anatomist. The first contact and union of the vena cava with the pulmonary veins, which occurs before the cava opens properly into the right ventricle of the heart, or gives off the coronary vein, a little above its escape from the liver, is by a lateral anastomosis; this is an ample foramen, of an oval form, communicating between the cava and the pulmonary vein, so that the blood is free to flow in the greatest abundance by that foramen from the vena cava into the pulmonary vein, and left auricle, and from thence into the left ventricle.

Farther, in this foramen ovale, from that part which regards the pulmonary vein, there is a thin tough membrane, larger than the opening, extended like an operculum or cover; this membrane in the adult blocking up the foramen, and adhering on all sides, finally closes it up, and almost obliterates every trace of it. In the fœtus, however, this membrane is so contrived that falling loosely upon itself, it permits a ready access to the lungs and heart, yielding a passage to the blood which is streaming from the cava, and hindering the tide at the same time from flowing back into that vein. All things, in short, permit us to believe that in the embryo the blood must constantly pass by this foramen from the vena cava into the pulmonary vein, and from thence into the left auricle of the heart; and having once entered there, it can never regurgitate.

Another union is that by the pulmonary artery, and is effected when that vessel divides into two branches after its escape from the right ventricle of the heart. It is as if to the two trunks already mentioned a third were superadded, a kind of arterial canal, carried obliquely from the pulmonary artery, to perforate and terminate in the great artery or aorta. So that in the dissection of the embryo, as it were, two aortas, or two roots of the great artery, appear springing from the heart. This canal shrinks gradually after birth, and after a time becomes withered, and finally almost removed, like the umbilical vessels.

The arterial canal contains no membrane or valve to direct or impede the flow of blood in this or in that direction: for at the root of the pulmonary artery, of which the arterial canal is the continuation in the fœtus, there are three semilunar valves, which open from within outwards, and oppose no obstacle to the blood flowing in this direction or from the right ventricle into the pulmonary artery and aorta; but they prevent all regurgitation from the aorta or pulmonic vessels back upon the right ventricle; closing with perfect accuracy, they oppose an effectual obstacle to everything of the kind in the embryo. So that there is also reason to believe that when the heart contracts, the blood is regularly propelled by the canal or passage indicated from the right ventricle into the aorta.

What is commonly said in regard to these two great communications, to wit, that they exist for the nutrition of the lungs, is both improbable and inconsistent; seeing that in the adult they are closed up, abolished, and consolidated, although the lungs, by reason of their heat and motion, must then be presumed to require a larger supply of nourishment. The same may be said in regard to the assertion that the heart in the embryo does not pulsate, that it neither acts nor moves, so that nature was forced to make these communications for the nutrition of the lungs. This is plainly false; for simple inspection of the incubated egg, and of embryos just taken out of the uterus, shows that the heart moves in them precisely as in adults, and that nature feels no such necessity. I have myself repeatedly seen these motions, and Aristotle is likewise witness of their reality. "The pulse," he observes, "inheres in the very constitution of the heart, and appears from the beginning as is learned both from the dissection of living animals and the formation of the chick in the egg."[1] But we further observe that the passages in question are not only pervious up to the period of birth in man, as well as in other animals, as anatomists in general have described them, but for several months subsequently, in some indeed for several years, not to say for the whole course of life; as, for example, in the goose, snipe, and various birds and many of the smaller animals. And this circumstance it was, perhaps, that imposed upon Botallus, who thought he had discovered a new passage for the blood from the vena cava into the left ventricle of the heart; and I own that when I met with the same arrangement in one of the larger members of the mouse family, in the adult state, I was myself at first led to something of a like conclusion.

From this it will be understood that in the human embryo, and in the embryos of animals in which the communications are not closed, the same thing happens, namely, that the heart by its motion propels the blood by obvious and open passages from the vena cava into the aorta through the cavities of both the ventricles, the right one receiving the blood from the auricle, and propelling it by the pulmonary

[1] Lib. de Spiritu, cap. v.

artery and its continuation, named the ductus arteriosus, into the aorta; the left, in like manner, charged by the contraction of its auricle, which has received its supply through the foramen ovale from the vena cava, contracting, and projecting the blood through the root of the aorta into the trunk of that vessel.

In embryos, consequently, whilst the lungs are yet in a state of inaction, performing no function, subject to no motion any more than if they had not been present, nature uses the two ventricles of the heart as if they formed but one, for the transmission of the blood. The condition of the embryos of those animals which have lungs, whilst these organs are yet in abeyance and not employed, is the same as that of those animals which have no lungs.

So it clearly appears in the case of the fœtus that the heart by its action transfers the blood from the vena cava into the aorta, and that by a route as obvious and open, as if in the adult the two ventricles were made to communicate by the removal of their septum. We therefore find that in the greater number of animals—in all, indeed, at a certain period of their existence—the channels for the transmission of the blood through the heart are conspicuous. But we have to inquire why in some creatures—those, namely, that have warm blood, and that have attained to the adult age, man among the number—we should not conclude that the same thing is accomplished through the substance of the lungs, which in the embryo, and at a time when the function of these organs is in abeyance, nature effects by the direct passages described, and which, indeed, she seems compelled to adopt through want of a passage by the lungs; or why it should be better (for nature always does that which is best) that she should close up the various open routes which she had formerly made use of in the embryo and fœtus, and still uses in all other animals. Not only does she thereby open up no new apparent channels for the passages of the blood, but she even shuts up those which formerly existed.

And now the discussion is brought to this point, that they who inquire into the ways by which the blood reaches the left ventricle of the heart and pulmonary veins from the vena cava, will pursue the wisest course if they seek by dissection to discover the causes why in the larger and more perfect animals of mature age nature has rather chosen to make the blood percolate the parenchyma of the lungs, than, as in other instances, chosen a direct and obvious course—for I assume that no other path or mode of transit can be entertained. It must be because the larger and more perfect animals are warmer, and when adult their heat greater—ignited, as I might say, and requiring to be damped or mitigated, that the blood is sent through the lungs, in order that it may be tempered by the air that is inspired, and prevented from boiling up, and so becoming extinguished, or something else of the sort. But to determine these matters, and explain them satisfactorily, were to enter on a speculation in regard to the office of the lungs and the ends for which they exist. Upon such a subject, as well as upon what pertains to respiration, to the necessity and use of the air, etc., as also to the variety and diversity of organs that exist in the bodies of animals in connexion with these matters, although I have made a vast number of observations, I shall not speak till I can more conveniently set them forth in a treatise apart, lest I should be held as wandering too wide of my present purpose, which is the use and motion of the heart, and be charged with speaking of things beside the question, and rather complicating and quitting than illustrating it. And now returning to my immediate subject, I go on with what yet remains for demonstration, viz., that in the more perfect and warmer adult animals, and man, the blood passes from the right ventricle of the heart by the pulmonary artery, into the lungs, and thence by the pulmonary veins into the left auricle, and from there into the left ventricle of the heart. And, first, I shall show that this may be so, and then I shall prove that it is so in fact.

CHAPTER VII

The Blood Passes through the Substance of the Lungs from the Right Ventricle of the Heart into the Pulmonary Veins and Left Ventricle

That this is possible, and that there is nothing to prevent it from being so, appears when we reflect on the way in which water permeating the earth produces springs and rivulets, or when we speculate on the means by which the sweat passes through the skin, or the urine through the substance of the kidneys. It is well known that persons who use the Spa waters or those of La Madonna, in the territories of Padua, or others of an acidulous or vitriolated nature, or who simply swallow drinks by the gallon, pass all off again within an hour or two by the bladder. Such a quantity of liquid must take some short time in the concoction: it must pass through the liver (it is allowed by all that the juices of the food we consume pass twice through this organ in the course of the day); it must flow through the veins, through the tissues of the kidneys, and through the ureters into the bladder.

To those, therefore, whom I hear denying that the blood, aye, the whole mass of the blood, may pass through the substance of the lungs, even as the nutritive juices percolate the liver, asserting such a proposition to be impossible, and by no means to be entertained as credible, I reply, with the poet, that they are of that race of men who, when they will, assent full readily, and when they will not, by no manner of means; who, when their assent is wanted, fear, and when it is not, fear not to give it.

The substance of the liver is extremely dense, so is that of the kidney; the lungs, however, are of a much looser texture, and if compared with the kidneys are absolutely spongy. In the liver there is no forcing, no impelling power; in the lungs the blood is forced on by the pulse of the right ventricle, the necessary effect of whose impulse is the distension of the vessels and the pores of the lungs. And then the lungs, in respiration, are perpetually rising and falling: motions, the effect of which must needs be to open and shut the pores and vessels, precisely as in the case of a sponge, and of parts having a spongy structure, when they are alternately compressed and again are suffered to expand. The liver, on the contrary, remains at rest, and is never seen to be dilated or constricted. Lastly, if no one denies the possibility in man, oxen, and the larger animals generally, of the whole of the ingested juices passing through the liver, in order to reach the vena cava, for this reason, that if nourishment is to go on, these juices must needs get into the veins, and there is no other way but the one indicated, why should not the same arguments be held of avail for the passage of the blood in adults through the lungs? Why not maintain, with Columbus, that skilful and learned anatomist, that it must be so from the capacity and structure of the pulmonary vessels, and from the fact of the pulmonary veins and ventricle corresponding with them, being always found to contain blood, which must needs have come from the veins, and by no other passage save through the lungs? Columbus, and we also, from what precedes, from dissections, and other arguments, conceive the thing to be clear. But as there are some who admit nothing unless upon authority, let them learn that the truth I am contending for can be confirmed from Galen's own words, namely, that not only may the blood be transmitted from the pulmonary artery into the pulmonary veins, then into the left ventricle of the heart, and from thence into the arteries of the body, but that this is effected by the ceaseless pulsation of the heart and the motion of the lungs in breathing.

There are, as everyone knows, three sigmoid or semilunar valves situated at the orifice of the pulmonary artery, which effectually prevent the blood sent into the vessel from returning into the cavity of the heart. Now Galen, explaining the use of these valves, and the necessity for them, employs the following language:[1] "There is everywhere a mutual anastomosis and inosculation of the arteries with the veins, and they severally transmit both blood and spirit, by certain invisible and undoubtedly very narrow passages. Now if the mouth of the pulmonary artery had stood in

[1] De Usu partium, lib. vi, cap. 10.

like manner continually open, and nature had found no contrivance for closing it when requisite, and opening it again, it would have been impossible that the blood could ever have passed by the invisible and delicate mouths, during the contractions of the thorax, into the arteries; for all things are not alike readily attracted or repelled; but that which is light is more readily drawn in, the instrument being dilated, and forced out again when it is contracted, than that which is heavy; and in like manner is anything drawn more rapidly along an ample conduit, and again driven forth, than it is through a narrow tube. But when the thorax is contracted the pulmonary veins, which are in the lungs, being driven inwardly, and powerfully compressed on every side, immediately force out some of the spirit they contain, and at the same time assume a certain portion of blood by those subtle mouths, a thing that could never come to pass were the blood at liberty to flow back into the heart through the great orifice of the pulmonary artery. But its return through this great opening being prevented, when it is compressed on every side, a certain portion of it distils into the pulmonary veins by the minute orifices mentioned." And shortly afterwards, in the next chapter, he says: "The more the thorax contracts, the more it strives to force out the blood, the more exactly do these membranes (viz., the semilunar valves) close up the mouth of the vessel, and suffer nothing to regurgitate." The same fact he has also alluded to in a preceding part of the tenth chapter: "Were there no valves, a three-fold inconvenience would result, so that the blood would then perform this lengthened course in vain; it would flow inwards during the disastoles of the lungs and fill all their arteries; but in the systoles, in the manner of the tide, it would ever and anon, like the Euripus, flow backwards and forwards by the same way, with a reciprocating motion, which would nowise suit the blood. This, however, may seem a matter of little moment: but if it meantime appear that the function of respiration suffer, then I think it would be looked upon as no trifle, etc." Shortly afterwards he says: "And then a third inconvenience, by no means to be thought lightly of, would follow, were the blood moved backwards during the expirations, had not our Maker instituted those supplementary membranes." In the eleventh chapter he concludes: "That they (the valves) have all a common use, and that it is to prevent regurgitation or backward motion; each, however, having a proper function, the one set drawing matters from the heart, and preventing their return, the other drawing matters into the heart, and preventing their escape from it. For nature never intended to distress the heart with needless labour, neither to bring aught into the organ which it had been better to have kept away, nor to take from it again aught which it was requisite should be brought. Since, then, there are four orifices in all, two in either ventricle, one of these induces, the other educes." And again he says: "Farther, since there is one vessel, which consists of a simple covering implanted in the heart, and another which is double, extending from it (Galen is here speaking of the right side of the heart, but I extend his observations to the left side also), a kind of reservoir had to be provided, to which both belonging, the blood should be drawn in by one, and sent out by the other."

Galen adduces this argument for the transit of the blood by the right ventricle from the vena cava into the lungs; but we can use it with still greater propriety, merely changing the terms, for the passage of the blood from the veins through the heart into the arteries. From Galen, however, that great man, that father of physicians, it clearly appears that the blood passes through the lungs from the pulmonary artery into the minute branches of the pulmonary veins, urged to this both by the pulses of the heart and by the motions of the lungs and thorax; that the heart, moreover, is incessantly receiving and expelling the blood by and from its ventricles, as from a magazine or cistern, and for this end it is furnished with four sets of valves, two serving for the induction and two for the eduction of the blood, lest, like the Euripus, it should be incommodiously sent hither and thither, or flow back into the cavity which it should have quitted, or quit the part where its presence was required, and so the heart might be oppressed with labour in vain, and the office of the lungs be interfered with.[2] Finally, our position that the blood is continually permeating from the right to the left ventricle, from the vena cava into the aorta, through the porosities of the lungs, plainly appears from this, that since the blood is incessantly sent from the right ventricle into the lungs by the pulmonary artery, and in like manner is incessantly drawn from the lungs into the left ventricle, as appears from what precedes and the position of the valves, it cannot do otherwise than pass through continuously. And then, as the blood is incessantly flowing into the right ventricle of the heart, and is continually passed out from the left, as appears in like manner, and as is obvious, both to sense and reason, it is impossible that the blood can do otherwise than pass continually from the vena cava into the aorta.

Dissection consequently shows distinctly what takes place in the majority of animals, and indeed in all, up to the period of their maturity; and that the same thing occurs in adults is equally certain, both from Galen's words, and what has already been said, only that in the former the transit is effected by open and obvious passages, in the latter by the hidden porosities of the lungs and the minute inosculations of vessels. It therefore appears that, although one ventricle of the heart, the left to wit, would suffice for the distribution of the blood over the body, and its eduction from the vena cava, as indeed is done in those creatures that have no lungs, nature, nevertheless, when she ordained that the same blood should also percolate the lungs, saw herself obliged to add the right ventricle, the pulse of which should force the blood from the vena cava through the lungs into the cavity of the left ventricle. In this way, it may be said, that the right ventricle is made for the sake of the lungs, and for the transmission of the blood through them, not for their nutrition; for it were unreasonable to suppose that the lungs should require so much more copious a supply of nutriment, and that of so much purer and more spirituous a nature as coming immediately from the ventricle of the heart, that either the brain, with its peculiarly pure substance, or

[2] See the Commentary of the learned Hofmann upon the Sixth Book of Galen, "De Usu partium," a work which I first saw after I had written what precedes.

the eyes, with their lustrous and truly admirable structure, or the flesh of the heart itself, which is more suitably nourished by the coronary artery.

CHAPTER VIII

Of the Quantity of Blood Passing Through the Heart from the Veins to the Arteries; and of the Circular Motion of the Blood

Thus far I have spoken of the passage of the blood from the veins into the arteries, and of the manner in which it is transmitted and distributed by the action of the heart; points to which some, moved either by the authority of Galen or Columbus, or the reasonings of others, will give in their adhesion. But what remains to be said upon the quantity and source of the blood which thus passes is of a character so novel and unheard-of that I not only fear injury to myself from the envy of a few, but I tremble lest I have mankind at large for my enemies, so much doth wont and custom become a second nature. Doctrine once sown strikes deep its root, and respect for antiquity influences all men. Still the die is cast, and my trust is in my love of truth and the candour of cultivated minds. And sooth to say, when I surveyed my mass of evidence, whether derived from vivisections, and my various reflections on them, or from the study of the ventricles of the heart and the vessels that enter into and issue from them, the symmetry and size of these conduits,—for nature doing nothing in vain, would never have given them so large a relative size without a purpose,—or from observing the arrangement and intimate structure of the valves in particular, and of the other parts of the heart in general, with many things besides, I frequently and seriously bethought me, and long revolved in my mind, what might be the quantity of blood which was transmitted, in how short a time its passage might be effected, and the like. But not finding it possible that this could be supplied by the juices of the ingested aliment without the veins on the one hand becoming drained, and the arteries on the other getting ruptured through the excessive charge of blood, unless the blood should somehow find its way from the arteries into the veins, and so return to the right side of the heart, I began to think whether there might not be a MOTION, AS IT WERE, IN A CIRCLE. Now, this I afterwards found to be true; and I finally saw that the blood, forced by the action of the left ventricle into the arteries, was distributed to the body at large, and its several parts, in the same manner as it is sent through the lungs, impelled by the right ventricle into the pulmonary artery, and that it then passed through the veins and along the vena cava, and so round to the left ventricle in the manner already indicated. This motion we may be allowed to call circular, in the same way as Aristotle says that the air and the rain emulate the circular motion of the superior bodies; for the moist earth, warmed by the sun, evaporates; the vapours drawn upwards are condensed, and descending in the form of rain, moisten the earth again. By this arrangement are generations of living things produced; and in like manner are tempests and meteors engendered by the circular motion, and by the approach and recession of the sun.

And similarly does it come to pass in the body, through the motion of the blood, that the various parts are nourished, cherished, quickened by the warmer, more perfect, vaporous, spirituous, and, as I may say, alimentive blood; which, on the other hand, owing to its contact with these parts, becomes cooled, coagulated, and so to speak effete. It then returns to its sovereign, the heart, as if to its source, or to the inmost home of the body, there to recover its state of excellence or perfection. Here it renews its fluidity, natural heat, and becomes powerful, fervid, a kind of treasury of life, and impregnated with spirits, it might be said with balsam. Thence it is again dispersed. All this depends on the motion and action of the heart.

The heart, consequently, is the beginning of life; the sun of the microcosm, even as the sun in his turn might well be designated the heart of the world; for it is the heart by whose virtue and pulse the blood is moved, perfected, and made nutrient, and is preserved from corruption and coagulation; it is the household divinity which, discharging its function, nourishes, cherishes, quickens the whole body, and is indeed the foundation of life, the source of all action. But of these things we shall speak more opportunely when we come to speculate upon the final cause of this motion of the heart.

As the blood-vessels, therefore, are the canals and agents that transport the blood, they are of two kinds, the cava and the aorta; and this not by reason of there being two sides of the body, as Aristotle has it, but because of the difference of office, not, as is commonly said, in consequence of any diversity of structure, for in many animals, as I have said, the vein does not differ from the artery in the thickness of its walls, but solely in virtue of their distinct functions and uses. A vein and an artery, both styled veins by the ancients, and that not without reason, as Galen has remarked, for the artery is the vessel which carries the blood from the heart to the body at large, the vein of the present day bringing it back from the general system to the heart; the former is the conduit from, the latter the channel to, the heart; the latter contains the cruder, effete blood, rendered unfit for nutrition; the former transmits the digested, perfect, peculiarly nutritive fluid.

CHAPTER IX

That there is a Circulation of the Blood is Confirmed from the First Proposition

But lest anyone should say that we give them words only, and make mere specious assertions without any foundation, and desire to innovate without sufficient cause, three points present themselves for confirmation, which, being stated, I conceive that the truth I contend for will follow necessarily, and appear as a thing obvious to all. First, the blood is incessantly transmitted by the action of the heart from the vena cava to the arteries in such quantity that it cannot be supplied from the ingesta, and in such a manner that the whole must very quickly pass through the organ; second, the blood under the influence of the arterial pulse enters and is impelled in a continuous, equable, and incessant stream through every part and member of the body, in much larger quantity than were sufficient for nutrition, or than the whole mass of fluids could supply; third, the veins in like manner return this blood incessantly to the heart from parts and members of the body. These points proved, I conceive it will be manifest that the blood circulates, revolves, propelled and then returning, from the heart to the extremities, from the extremities to the heart, and thus that it performs a kind of circular motion.

Let us assume, either arbitrarily or from experiment, the quantity of blood which the left ventricle of the heart will contain when distended, to be, say, two ounces, three ounces, or one ounce and a half—in the dead body I have found it to hold upwards of two ounces. Let us assume further how much less the heart will hold in the contracted than in the dilated state; and how much blood it will project into the aorta upon each contraction; and all the world allows that with the systole something is always projected, a necessary consequence demonstrated in the third chapter, and obvious from the structure of the valves; and let us suppose as approaching the truth that the fourth, or fifth, or sixth, or even but the eighth part of its charge is thrown into the artery at each contraction; this would give either half an ounce, or three drachms, or one drachm of blood as propelled by the heart at each pulse into the aorta; which quantity, by reason of the valves at the root of the vessel, can by no means return into the ventricle. Now, in the course of half an hour, the heart will have made more than one thousand beats, in some as many as two, three, and even four thousand. Multiplying the number of drachms propelled by the number of pulses, we shall have either one thousand half ounces, or one thousand times three drachms, or a like proportional quantity of blood, according to the amount which we assume as propelled with each stroke of the heart, sent from this organ into the artery—a larger quantity in every case than is contained in the whole body! In the same way, in the sheep or dog, say but a single scruple of blood passes with each stroke of the heart, in one half-hour we should have one thousand scruples, or about three pounds and a half, of blood injected into the aorta; but the body of neither animal contains above four pounds of blood, a fact which I have myself ascertained in the case of the sheep.

Upon this supposition, therefore, assumed merely as a ground for reasoning, we see the whole mass of blood passing through the heart, from the veins to the arteries, and in like manner through the lungs.

But let it be said that this does not take place in half an hour, but in an hour, or even in a day; any way, it is still manifest that more blood passes through the heart in consequence of its action, than can either be supplied by the whole of the ingesta, or than can be contained in the veins at the same moment.

Nor can it be allowed that the heart in contracting sometimes propels and sometimes does not propel, or at most propels but very little, a mere nothing, or an imaginary something: all this, indeed, has already been refuted, and is, besides, contrary both to sense and reason. For if it be a necessary effect of the dilatation of the heart that its ventricles become filled with blood, it is equally so that, contracting, these cavities should expel their contents; and this not in any trifling measure. For neither are the conduits small, nor the contractions few in number, but frequent, and always in some certain proportion, whether it be a third or a sixth, or an eighth, to the total capacity of the ventricles, so that a like proportion of blood must be expelled, and a like proportion received with each stroke of the heart, the capacity of the ventricle contracted always bearing a certain relation to the capacity of the ventricle when dilated. And since, in dilating, the ventricles cannot be supposed to get filled with nothing, or with an imaginary something, so in contracting they never expel nothing or aught imaginary, but always a certain something, viz., blood, in proportion to the amount of the contraction. Whence it is to be concluded that if at one stroke the heart of man, the ox, or the sheep, ejects but a single drachm of blood and there are one thousand strokes in half an hour, in this interval there will have been ten pounds five ounces expelled; if with each stroke two drachms are expelled, the quantity would, of course, amount to twenty pounds and ten ounces; if half an ounce, the quantity would come to forty-one pounds and eight ounces; and were there one ounce, it would be as much as eighty-three pounds and four ounces; the whole of which, in the course of one-half hour, would have been transfused from the veins to the arteries. The actual quantity of blood expelled at each stroke of the heart, and the circumstances under which it is either greater or less than ordinary, I leave for particular determination afterwards, from numerous observations which I have made on the subject.

Meantime this much I know, and would here proclaim to all, that the blood is transfused at one time in larger, at another in smaller, quantity; and that the circuit of the blood is accomplished now more rapidly, now more slowly, according to the temperament, age, etc., of the individual, to external and internal circumstances, to naturals and non-naturals—sleep, rest, food, exercise, affections of the mind, and the like. But, supposing even the smallest quantity of blood to be passed through the heart and the lungs with each pulsation, a vastly greater amount would still be thrown into the arteries and whole body than could by any possibility be supplied by the food consumed. It could be furnished in no other way than by making a circuit and returning.

This truth, indeed, presents itself obviously before us when we consider what happens in the dissection of living animals; the great artery need not be divided, but a very small branch only (as Galen even proves in regard to man), to have the whole of the blood in the body, as well that of the veins as of the arteries, drained away in the course of no long time—some half-hour or less. Butchers are well aware of the fact and can bear witness to it; for, cutting the throat of an ox and so dividing the vessels of the neck, in less than a quarter of an hour they have all the vessels bloodless—the whole mass of blood has escaped. The same thing also occasionally occurs with great rapidity in performing amputations and removing tumors in the human subject.

Nor would this argument lose of its force, did any one say that in killing animals in the shambles, and performing amputations, the blood escaped in equal, if not perchance in larger quantity by the veins than by the arteries. The contrary of this statement, indeed, is certainly the truth; the veins, in fact, collapsing, and being without any propelling power, and further, because of the impediment of the valves, as I shall show immediately, pour out but very little blood; whilst the arteries spout it forth with force abundantly, impetuously, and as if it were propelled by a syringe. And then the experiment is easily tried of leaving the vein untouched and only dividing the artery in the neck of a sheep or dog, when it will be seen with what force, in what abundance, and how quickly, the whole blood in the body, of the veins as well as of the arteries, is emptied. But the arteries receive blood from the veins in no other way than by transmission through the heart, as we have already seen; so that if the aorta be tied at the base of the heart, and the carotid or any other artery be opened, no one will now be surprised to find it empty, and the veins only replete with blood.

And now the cause is manifest, why in our dissections we usually find so large a quantity of blood in the veins, so little in the arteries; why there is much in the right ventricle, little in the left, which probably led the ancients to believe that the arteries (as their name implies) contained nothing but spirits during the life of an animal. The true cause of the difference is perhaps this, that as there is no passage to the arteries, save through the lungs and heart, when an animal has ceased to breathe and the lungs to move, the blood in the pulmonary artery is prevented from passing into the pulmonary veins, and from thence into the left ventricle of the heart; just as we have already seen the same transit prevented in the embryo, by the want of movement in the lungs and the alternate opening and shutting of their hidden and invisible porosities and apertures. But the heart not ceasing to act at the same precise moment as the lungs, but surviving them and continuing to pulsate for a time, the left ventricle and arteries go on distributing their blood to the body at large and sending it into the veins; receiving none from the lungs, however, they are soon exhausted, and left, as it were, empty. But even this fact confirms our views, in no trifling manner, seeing that it can be ascribed to no other than the cause we have just assumed.

Moreover, it appears from this that the more frequently or forcibly the arteries pulsate, the more speedily will the body be exhausted of its blood during hemorrhage. Hence, also, it happens, that in fainting fits and in states of alarm, when the heart beats more languidly and less forcibly, hemorrhages are diminished and arrested.

Still further, it is from this, that after death, when the heart has ceased to beat, it is impossible, by dividing either the jugular or femoral veins and arteries, by any effort, to force out more than one-half of the whole mass of the blood. Neither could the butchers ever bleed the carcass effectually did he neglect to cut the throat of the ox which he has knocked on the head and stunned, before the heart had ceased beating.

Finally, we are now in a condition to suspect wherefore it is that no one has yet said anything to the purpose upon the anastomosis of the veins and arteries, either as to where or how it is effected, or for what purpose. I now enter upon the investigation of the subject.

CHAPTER X

The First Position: of the Quantity of Blood Passing from the Veins to the Arteries. And that there is a Circuit of the Blood, Freed from Objections, and Farther Confirmed by Experiment

So far our first position is confirmed, whether the thing be referred to calculation or to experiment and dissection, viz., that the blood is incessantly poured into the arteries in larger quantities than it can be supplied by the food; so that the whole passing over in a short space of time, it is matter of necessity that the blood perform a circuit, that it return to whence it set out.

But if anyone shall here object that a large quantity may pass through and yet no necessity be found for a circulation, that all may come from the meat and drink consumed, and quote as an illustration the abundant supply of milk in the mammæ—for a cow will give three, four, and even seven gallons and more in a day, and a woman two or three pints whilst nursing a child or twins, which must manifestly be derived from the food consumed; it may be answered that the heart by computation does as much and more in the course of an hour or two.

And if not yet convinced, he shall still insist that when an artery is divided, a preternatural route is, as it were, opened, and that so the blood escapes in torrents, but that the same thing does not happen in the healthy and uninjured body when no outlet is made; and that in arteries filled, or in their natural state, so large a quantity of blood cannot pass in so short a space of time as to make any return necessary—to all this it may be answered that, from the calculation already made, and the reasons assigned, it appears that by so much as the heart in its dilated state contains, in addition to its contents in the state of constriction, so much in a general way must it emit upon each pulsation, and in such quantity must the blood pass, the body being entire and naturally constituted.

But in serpents, and several fishes, by tying the veins some way below the heart you will perceive a space between the ligature and the heart speedily to become empty; so that, unless you would deny the evidence of your senses, you must needs admit the return of the blood to the heart. The same thing will also plainly appear when we come to discuss our second position.

Let us here conclude with a single example, confirming all that has been said, and from which everyone may obtain conviction through the testimony of his own eyes.

If a live snake be laid open, the heart will be seen pulsating quietly, distinctly, for more than an hour, moving like a worm, contracting in its longitudinal dimensions, (for it is of an oblong shape), and propelling its contents. It becomes of a paler colour in the systole, of a deeper tint in the diastole; and almost all things else are seen by which I have already said that the truth I contend for is established, only that here everything takes place more slowly, and is more distinct. This point in particular may be observed more clearly than the noonday sun: the vena cava enters the heart at its lower part, the artery quits it at the superior part; the vein being now seized either with forceps or between the finger and the thumb, and the course of the blood for some space below the heart interrupted, you will perceive the part that intervenes between the fingers and the heart almost immediately to become empty, the blood being exhausted by the action of the heart; at the same time the heart will become of a much paler colour, even in its state of dilatation, than it was before; it is also smaller than at first, from wanting blood: and then it begins to beat more slowly, so that it seems at length as if it were about to die. But the impediment to the flow of blood being removed, instantly the colour and the size of the heart are restored.

If, on the contrary, the artery instead of the vein be compressed or tied, you will observe the part between the obstacle and the heart, and the heart itself, to become inordinately distended, to assume a deep purple or even livid colour, and at length to be so much oppressed with blood that you will believe it about to be choked; but the obstacle removed, all things immediately return to their natural state and colour, size, and impulse.

Here then we have evidence of two kinds of death: extinction from deficiency, and suffocation from excess. Examples of both have now been set before you, and you have had opportunity of viewing the truth contended for with your own eyes in the heart.

CHAPTER XI

The Second Position is Demonstrated

That this may the more clearly appear to everyone, I have here to cite certain experiments, from which it seems obvious that the blood enters a limb by the arteries, and returns from it by the veins; that the arteries are the vessels carrying the blood from the heart, and the veins the returning channels of the blood to the heart; that in the limbs and extreme parts of the body the blood passes either immediately by anastomosis from the arteries into the veins, or mediately by the porosities of the flesh, or in both ways, as has already been said in speaking of the passage of the blood through the lungs whence it appears manifest that in the circuit the blood moves from that place to this place, and from that point to this one; from the centre to the extremities, to wit; and from the extreme parts back to the centre. Finally, upon grounds of calculation, with the same elements as before, it will be obvious that the quantity can neither be accounted for by the ingesta, nor yet be held necessary to nutrition.

The same thing will also appear in regard to ligatures, and wherefore they are said to *draw;* though this is neither from the heat, nor the pain, nor the vacuum they occasion, nor indeed from any other cause yet thought of; it will also explain the uses and advantages to be derived from ligatures in medicine, the principle upon which they either suppress or occasion hemorrhage; how they induce sloughing and more extensive mortification in extremities; and how they act in the castration of animals and the removal of warts and fleshy tumours. But it has come to pass, from no one having duly weighed and understood the cause and rationale of these various effects, that though almost all, upon the faith of the old writers, recommend ligatures in the treatment of disease, yet very few comprehend their proper employment, or derive any real assistance from them in effecting cures.

Ligatures are either very tight or of medium tightness. A ligature I designate as tight or perfect when it so constricts an extremity that no vessel can be felt pulsating beyond it. Such a ligature we use in amputations to control the flow of blood; and such also are employed in the castration of animals and the ablation of tumours. In the latter instances, all afflux of nutriment and heat being prevented by the ligature, we see the testes and large fleshy tumours dwindle, die, and finally fall off.

Ligatures of medium tightness I regard as those which compress a limb firmly all round, but short of pain, and in such a way as still suffers a certain degree of pulsation to be felt in the artery beyond them. Such a ligature is in use in blood-letting, an operation in which the fillet applied above the elbow is not drawn so tight but that the arteries at the wrist may still be felt beating under the finger.

Now let anyone make an experiment upon the arm of a man, either using such a fillet as is employed in blood-letting, or grasping the limb lightly with his hand, the best subject for it being one who is lean, and who has large veins, and the best time after exercise, when the body is warm, the pulse is full, and the blood carried in larger quantity to the extremities, for all then is more conspicuous; under such circumstances let a ligature be thrown about the extremity, and drawn as tightly as can be borne, it will first be perceived that beyond the ligature, neither in the wrist nor anywhere else, do the arteries pulsate, at the same time that immediately above the ligature the artery begins to rise higher at each diastole, to throb more violently, and to swell in its vicinity with a kind of tide, as if it strove to break through and overcome the obstacle to its current; the artery here, in short, appears as if it were preternaturally full. The hand under such circumstances retains its natural colour and appearance; in the course of time it begins to fall somewhat in temperature, indeed, but nothing is *drawn* into it.

After the bandage has been kept on for some short time in this way, let it be slackened a little, brought to that state or term of medium tightness which is used in bleeding, and it will be seen that the whole hand and arm will instantly become deeply coloured and distended, and the veins show themselves tumid and knotted; after ten or twelve pulses of the artery, the hand will be perceived excessively distended, injected, gorged with blood, *drawn*, as it is said, by this medium ligature, without pain, or heat, or any horror of a vacuum, or any other cause yet indicated.

If the finger be applied over the artery as it is pulsating by the edge of the fillet, at the moment of slackening it, the blood will be felt to glide through, as it were, underneath the finger; and he, too, upon whose arm the experiment is made, when the ligature is slackened, is distinctly conscious of a sensation of warmth, and of something, viz., a stream of blood suddenly making its way along the course of the vessels and diffusing itself through the hand, which at the same time begins to feel hot, and becomes distended.

As we had noted, in connexion with the tight ligature, that the artery above the bandage was distended and pulsated, not below it, so, in the case of the moderately tight bandage, on the contrary, do we find that the veins below, never above, the fillet, swell, and become dilated, whilst the arteries shrink; and such is the degree of distension of the veins here, that it is only very strong pressure that will force the blood beyond the fillet, and cause any of the veins in the upper part of the arm to rise.

From these facts it is easy for every careful observer to learn that the blood enters an extremity by the arteries; for when they are effectually compressed nothing is *drawn* to the member; the hand preserves its colour; nothing flows into it, neither is it distended; but when the pressure is diminished, as it is with the bleeding fillet, it is manifest that the blood is instantly thrown in with force, for then the hand begins to swell; which is as much as to say, that when the arteries pulsate the blood is flowing through them, as it is when the moderately tight ligature is applied; but where they do not pulsate, as, when a tight ligature is used, they cease from transmitting anything, they are only distended above the part where the ligature is applied. The veins again being compressed, nothing can flow through them; the certain indication of which is, that below the ligature they are much more tumid than above it, and than they usually appear when there is no bandage upon the arm.

It therefore plainly appears that the ligature prevents the return of the blood through the veins to the parts above it, and maintains those beneath it in a state of permanent distension. But the arteries, in spite of its pressure, and under the force and impulse of the heart, send on the blood from the internal parts of the body to the parts beyond the ligature. And herein consists the difference between the tight and the medium ligature, that the former not only prevents the passage of the blood in the veins, but in the arteries also; the latter, however, whilst it does not prevent the force of the pulse from extending beyond it, and so propelling the blood to the extremities of the body, compresses the veins, and greatly or altogether impedes the return of the blood through them.

Seeing, therefore, that the moderately tight ligature renders the veins turgid and distended, and the whole hand full of blood, I ask, whence is this? Does the blood accumulate below the ligature coming through the veins, or through the arteries, or passing by certain hidden porosities? Through the veins it cannot come; still less can it come through invisible channels; it must needs, then, arrive by the arteries, in conformity with all that has been already said. That it cannot flow in by the veins appears plainly enough from the fact that the blood cannot be forced towards the heart unless the ligature be removed; when this is done suddenly all the veins collapse, and disgorge themselves of their contents into the superior parts, the hand at the same time resumes its natural pale colour, the tumefaction and the stagnating blood having disappeared.

Moreover, he whose arm or wrist has thus been bound for some little time with the medium bandage, so that it has not only got swollen and livid but cold, when the fillet is undone is aware of something cold making its way upwards along with the returning blood, and reaching the elbow or the axilla. And I have myself been inclined to think that this cold blood rising upwards to the heart was the cause of the fainting that often occurs after blood-letting: fainting frequently supervenes even in robust subjects, and mostly at the moment of undoing the fillet, as the vulgar say, from the turning of the blood.

Farther, when we see the veins below the ligature instantly swell up and become gorged, when from extreme tightness it is somewhat relaxed, the arteries meantime continuing unaffected, this is an obvious indication that the blood passes from the arteries into the veins, and not from the veins into the arteries, and that there is either an anastomosis of the two orders of vessels, or porosities in the flesh and solid parts generally that are permeable to the blood. It is farther an indication that the veins have frequent communications with one another, because they all become turgid together, whilst under the medium ligature applied above the elbow; and if any single small vein be pricked with a lancet, they all speedily shrink, and disburthening themselves into this they subside almost simultaneously.

These considerations will enable anyone to understand the nature of the attraction that is exerted by ligatures, and perchance of fluxes generally; how, for example, when the veins are compressed by a bandage of medium tightness applied above the elbow, the blood cannot escape, whilst it still continues to be driven in, by the forcing power of the heart, by which the parts are of necessity filled, gorged with blood. And how should it be otherwise? Heat and pain and a vacuum draw, indeed; but in such wise only that parts are filled, not preternaturally distended or gorged, and not so suddenly and violently overwhelmed with the charge of blood forced in upon them, that the flesh is lacerated and the vessels ruptured. Nothing of the kind as an effect of heat, or pain, or the vacuum force, is either credible or demonstrable.

Besides, the ligature is competent to occasion the afflux in question without either pain, or heat, or a vacuum. Were pain in any way the cause, how should it happen that, with the arm bound above the elbow, the hand and fingers should swell below the bandage, and their veins become distended? The pressure of the bandage certainly prevents the blood from getting there by the veins. And then, wherefore is there neither swelling nor repletion of the veins, nor any sign or symptom of attraction or afflux, above the ligature? But this is the obvious cause of the preternatural attraction and swelling below the bandage, and in the hand and fingers, that the blood is entering abundantly, and with force, but cannot pass out again.

Now is not this the cause of all tumefaction, as indeed Avicenna has it, and of all oppressive redundancy in parts, that the access to them is open, but the egress from them is closed? Whence it comes that they are gorged and tumefied. And may not the same thing happen in local inflammations, where, so long as the swelling is on the increase, and has not reached its extreme term, a full pulse is felt in the part, especially when the disease is of the more acute kind, and the swelling usually takes place most rapidly. But these are matters for after discussion. Or does this, which occurred in my own case, happen from the same cause? Thrown from a carriage upon one occasion, I struck my forehead a blow upon the place where a twig of the artery advances from the temple, and immediately, within the time in which twenty beats could have been made I felt a tumour the size of an egg developed, without either heat or any great pain: the near vicinity of the artery had caused the blood to be effused into the bruised part with unusual force and velocity.

And now, too, we understand why in phlebotomy we apply our ligature above the part that is punctured, not below it; did the flow come from above, not from below, the constriction in this case would not only be of no service, but would prove a positive hindrance; it would have to be applied below the orifice, in order to have the flow more free, did the blood descend by the veins from superior to inferior parts; but as it is elsewhere forced through the extreme arteries into the extreme veins, and the return in these last is opposed by the ligature, so do they fill and swell, and being thus filled and distended, they are made capable of projecting their charge with force, and to a distance, when any one of them is suddenly punctured; but the ligature being slackened, and the returning channels thus left open, the blood forthwith no longer escapes, save by drops; and, as all the world knows, if in performing phlebotomy the bandage be either slackened too much or the limb be bound too tightly, the blood escapes without force, because in the one case the returning channels are not adequately obstructed; in the other the channels of influx, the arteries, are impeded.

CHAPTER XII

That there is a Circulation of the Blood is Shown from the Second Position Demonstrated

If these things be so, another point which I have already referred to, viz., the continual passage of the blood through the heart will also be confirmed. We have seen, that the blood passes from the arteries into the veins, not from the veins into the arteries; we have seen, farther, that almost the whole of the blood may be withdrawn from a puncture made in one of the cutaneous veins of the arm if a bandage properly applied be used; we have seen, still farther, that the blood flows so freely and rapidly that not only is the whole quantity which was contained in the arm beyond the ligature, and before the puncture was made, discharged, but the whole which is contained in the body, both that of the arteries and that of the veins.

Whence we must admit, first, that the blood is sent along with an impulse, and that it is urged with force below the ligature; for it escapes with force, which force it receives from the pulse and power of the heart; for the force and motion of the blood are derived from the heart alone. Second, that the afflux proceeds from the heart, and through the heart by a course from the great veins; for it gets into the parts below the ligature through the arteries, not through the veins; and the arteries nowhere receive blood from the veins, nowhere receive blood save and except from the left ventricle of the heart. Nor could so large a quantity of blood be drawn from one vein (a ligature having been duly applied), nor with such impetuousity, such readiness, such celerity, unless through the medium of the impelling power of the heart.

But if all things be as they are now represented, we shall feel ourselves at liberty to calculate the quantity of the blood, and to reason on its circular motion. Should anyone, for instance, performing phlebotomy, suffer the blood to flow in the manner it usually does, with force and freely, for some half hour or so, no question but that the greatest part of the blood being abstracted, faintings and syncopes would ensue, and that not only would the arteries but the great veins also be nearly emptied of their contents. It is only consonant with reason to conclude that in the course of the half hour hinted at, so much as has escaped has also passed from the great veins through the heart into the aorta. And further, if we calculate how many ounces flow through one arm, or how many pass in twenty or thirty pulsations under the medium ligature, we shall have some grounds for estimating how much passes through the other arm in the same space of time: how much through both lower extremities, how much through the neck on either side, and through all the other arteries and veins of the body, all of which have been supplied with fresh blood, and as this blood must have passed through the lungs and ventricles of the heart, and must have come from the great veins,—we shall perceive that a circulation is absolutely necessary, seeing that the quantities hinted at cannot be supplied immediately from the ingesta, and are vastly more than can be requisite for the mere nutrition of the parts.

It is still further to be observed, that in practising phlebotomy the truths contended for are sometimes confirmed in another way; for having tied up the arm properly, and made the puncture duly, still, if from alarm or any other causes, a state of faintness supervenes, in which the heart always pulsates more languidly, the blood does not flow freely, but distils by drops only. The reason is, that with a somewhat greater than usual resistance offered to the transit of the blood by the bandage, coupled with the weaker action of the heart, and its diminished impelling power, the stream cannot make its way under the ligature; and farther, owing to the weak and languishing state of the heart, the blood is not transferred in such quantity as wont from the veins to the arteries through the sinuses of that organ. So also, and for the same reasons, are the menstrual fluxes of women, and indeed hemorrhages of every kind, controlled. And now, a contrary state of things occurring, the patient getting rid of his fear and recovering his courage, the pulse strength is increased, the arteries begin again to beat with greater force, and to drive the blood even into the part that is bound; so that the blood now springs from the puncture in the vein, and flows in a continuous stream.

CHAPTER XIII

The Third Position is Confirmed: and the Circulation of the Blood is Demonstrated from It

Thus far we have spoken of the quantity of blood passing through the heart and the lungs in the centre of the body, and in like manner from the arteries into the veins in the peripheral parts and the body at large. We have yet to explain, however, in what manner the blood finds its way back to the heart from the extremities by the veins, and how and in what way these are the only vessels that convey the blood from the external to the central parts; which done, I conceive that the three fundamental propositions laid down for the circulation of the blood will be so plain, so well established, so obviously true, that they may claim general credence. Now the remaining position will be made sufficiently clear from the valves which are found in the cavities of the veins themselves, from the uses of these, and from experiments cognizable by the senses.

The celebrated Hieronymus Fabricius of Aquapendente, a most skilful anatomist, and venerable old man, or, as the learned Riolan will have it, Jacobus Silvius, first gave representations of the valves in the veins, which consist of raised or loose portions of the inner membranes of these vessels, of extreme delicacy, and a sigmoid or semilunar shape. They are situated at different distances from one another, and diversely in different individuals; they are connate at the sides of the veins; they are directed upwards towards the trunks of the veins; the two—for there are for the most part two together—regard each other, mutually touch, and are so ready to come into contact by their edges, that if anything attempts to pass from the trunks into the branches of the veins, or from the greater vessels into the less, they completely prevent it; they are farther so arranged, that the horns of those that succeed are opposite the middle of the convexity of those that precede, and so on alternately.

The discoverer of these valves did not rightly understand their use, nor have succeeding anatomists added anything to our knowledge: for their office is by no means explained when we are told that it is to hinder the blood, by its weight, from all flowing into inferior parts; for the edges of the valves in the jugular veins hang downwards, and are so contrived that they prevent the blood from rising upwards; the valves, in a word, do not invariably look upwards, but always toward the trunks of the veins, invariably towards the seat of the heart. I, and indeed others, have sometimes found valves in the emulgent veins, and in those of the mesentery, the edges of which were directed towards the vena cava and vena portæ. Let it be added that there are no valves in the arteries, and that dogs, oxen, etc., have invariably valves at the divisions of their crural veins, in the veins that meet towards the top of the os sacrum, and in those branches which come from the haunches, in which no such effect of gravity from the erect position was to be apprehended. Neither are there valves in the jugular veins for the purpose of guarding against apoplexy, as some have said; because in sleep the head is more apt to be influenced by the contents of the carotid arteries. Neither are the valves present, in order that the blood may be retained in the divarications or smaller trunks and minuter branches, and not be suffered to flow entirely into the more open and capacious channels; for they occur where there are no divarications; although it must be owned that they are most frequent at the points where branches join. Neither do they exist for the purpose of rendering the current of blood more slow from the centre of the body; for it seems likely that the blood would be disposed to flow with sufficient slowness of its own accord, as it would have to pass from larger into continually smaller vessels, being separated from the mass and fountain head, and attaining from warmer into colder places.

But the valves are solely made and instituted lest the blood should pass from the greater into the lesser veins, and either rupture them or cause them to become varicose; lest, instead of advancing from the extreme to the central parts of the body, the blood should rather proceed along the veins from the centre to the extremities; but the delicate valves, while they readily open in the right direction, entirely prevent all such contrary motion, being so situated and arranged, that if anything escapes, or is less perfectly obstructed by the cornua of the one above, the fluid passing, as it were, by the chinks between the cornua, it is immediately received on the convexity of the one beneath, which is placed transversely with reference to the former, and so is effectually hindered from getting any farther.

And this I have frequently experienced in my dissections of the veins: if I attempted to pass a probe from the trunk of the veins into one of the smaller branches, whatever care I took I found it impossible to introduce it far any way, by reason of the valves; whilst, on the contrary, it was most easy to push it along in the opposite direction, from without inwards, or from the branches towards the trunks and roots. In many places two valves are so placed and fitted, that when raised they come exactly together in the middle of the vein, and are there united by the contact of their margins; and so accurate is the adaptation, that neither by the eye nor by any other means of examination, can the slightest chink along the line of contact be perceived. But if the probe be now introduced from the extreme towards the more central parts, the valves, like the floodgates of a river, give way, and are most readily pushed aside. The effect of this arrangement plainly is to prevent all motion of the blood from the heart and vena cava, whether it be upwards towards the head, or downwards towards the feet, or to either side towards the arms, not a drop can pass: all motion of the blood. beginning in the larger and tending towards the smaller veins, is opposed and resisted by them; whilst the motion that proceeds from the lesser to end in the larger branches is favoured, or, at all events, a free and open passage is left for it.

But that this truth may be made the more apparent, let an arm be tied up above the elbow as if for phlebotomy (A, A, fig. 1). At intervals in the course of the veins, especially in labouring people and those whose veins are large, certain knots or elevations (B, C, D, E, F) will be perceived, and this not only at the places where a branch is received (E. F), but also where none enters (C, D): these knots or risings are all formed by valves, which thus show themselves externally. And now if you press the blood from the space above one of the valves, from H to O, (fig. 2,) and keep the point of a finger upon the vein inferiorly, you will see no influx of blood from above; the portion of the vein between the point of the finger and the valve O will be obliterated; yet will the vessel continue sufficiently distended above the valve (O, G). The blood being thus pressed out and the vein emptied, if you now apply a finger of the other hand upon the distended part of the vein above the valve O, (fig. 3,) and press downwards, you will find that you cannot force the blood through or beyond the valve; but the greater effort you use, you will only see the portion of vein that is between the finger and the valve become more distended, that portion of the vein which is below the valve remaining all the while empty (H, O, fig. 3).

It would therefore appear that the function of the valves in the veins is the same as that of the three sigmoid valves which we find at the commencement of the aorta and pulmonary artery, viz., to prevent all reflux of the blood that is passing over them.

[NOTE.—Woodcuts of the veins of the arm to which these letters and figures refer appear here in the original.—C. N. B. C.]

Farther, the arm being bound as before, and the veins looking full and distended, if you press at one part in the course of a vein with the point of a finger (L, fig. 4), and then with another finger streak the blood upwards beyond the next valve (N), you will perceive that this portion of the vein continues empty (L. N), and that the blood cannot retrograde, precisely as we have already seen the case to be in fig. 2; but the finger first applied (H, fig. 2, L, fig. 4), being removed, immediately the vein is filled from below, and the arm becomes as it appears at D C, fig. 1. That the blood in the veins therefore proceeds from inferior or more remote parts, and towards the heart, moving in these vessels in this and not in the contrary direction, appears most obviously. And although in some places the valves, by not acting with such perfect accuracy, or where there is but a single valve, do not seem totally to prevent the passage of the blood from the centre, still the greater number of them plainly do so; and then, where things appear contrived more negligently, this is compensated either by the more frequent occurrence or more perfect action of the succeeding valves, or in some other way: the veins in short, as they are the free and open conduits of the blood returning *to* the heart, so are they effectually prevented from serving as its channels of distribution *from* the heart.

But this other circumstance has to be noted: The arm being bound, and the veins made turgid, and the valves prominent, as before, apply the thumb or finger over a vein in the situation of one of the valves in such a way as to compress it, and prevent any blood from passing upwards from the hand; then, with a finger of the other hand, streak the blood in the vein upwards till it has passed the next valve above (N, fig. 4), the vessel now remains empty; but the finger at L being removed for an instant, the vein is immediately filled from below; apply the finger again, and having in the same manner streaked the blood upwards, again remove the finger below, and again the vessel becomes distended as before; and this repeat, say a thousand times, in a short space of time. And now compute the quantity of blood which you have thus pressed up beyond the valve, and then multiplying the assumed quantity by one thousand, you will find that so much blood has passed through a certain portion of the vessel; and I do now believe that you will find yourself convinced of the circulation of the blood, and of its rapid motion. But if in this experiment you say that a violence is done to nature, I do not doubt but that, if you proceed in the same way, only taking as great a length of vein as possible, and merely remark with what rapidity the blood flows upwards, and fills the vessel from below, you will come to the same conclusion.

CHAPTER XIV

Conclusion of the Demonstration of the Circulation

And now I may be allowed to give in brief my view of the circulation of the blood, and to propose it for general adoption.

Since all things, both argument and ocular demonstration, show that the blood passes through the lungs, and heart by the force of the ventricles, and is sent for distribution to all parts of the body, where it makes its way into the veins and porosities of the flesh, and then flows by the veins from the circumference on every side to the centre, from the lesser to the greater veins, and is by them finally discharged into the vena cava and right auricle of the heart, and this in such a quantity or in such a flux and reflux thither by the arteries, hither by the veins, as cannot possibly be supplied by the ingesta, and is much greater than can be required for mere purposes of nutrition; it is absolutely necessary to conclude that the blood in the animal body is impelled in a circle, and is in a state of ceaseless motion; that this is the act or function which the heart performs by means of its pulse; and that it is the sole and only end of the motion and contraction of the heart.

CHAPTER XV

The Circulation of the Blood is Further Confirmed by Probable Reasons

It will not be foreign to the subject if I here show further, from certain familiar reasonings, that the circulation is matter both of convenience and necessity. In the first place, since death is a corruption which takes place through deficiency of heat,[1] and since all living things are warm, all dying things cold, there must be a particular seat and fountain, a kind of home and hearth, where the cherisher of nature, the original of the native fire, is stored and preserved; from which heat and life are dispensed to all

[1] Aristoteles De Respiratione, lib. ii et iii: De Part. Animal. et alibi.

parts as from a fountain head; from which sustenance may be derived; and upon which concoction and nutrition, and all vegetative energy may depend. Now, that the heart is this place, that the heart is the principle of life, and that all passes in the manner just mentioned, I trust no one will deny.

The blood, therefore, required to have motion, and indeed such a motion that it should return again to the heart; for sent to the external parts of the body far from its fountain, as Aristotle says, and without motion it would become congealed. For we see motion generating and keeping up heat and spirits under all circumstances, and rest allowing them to escape and be dissipated. The blood, therefore, becoming thick or congealed by the cold of the extreme and outward parts, and robbed of its spirits, just as it is in the dead, it was imperative that from its fount and origin, it should again receive heat and spirits, and all else requisite to its preservation—that, by returning, it should be renovated and restored.

We frequently see how the extremities are chilled by the external cold, how the nose and cheeks and hands look blue, and how the blood, stagnating in them as in the pendent or lower parts of a corpse, becomes of a dusky hue; the limbs at the same time getting torpid, so that they can scarcely be moved, and seem almost to have lost their vitality. Now they can by no means be so effectually, and especially so speedily restored to heat and colour and life, as by a new efflux and contact of heat from its source. But how can parts attract in which the heat and life are almost extinct? Or how should they whose passages are filled with condensed and frigid blood, admit fresh aliment—renovated blood—unless they had first got rid of their old contents? Unless the heart were truly that fountain where life and heat are restored to the refrigerated fluid, and whence new blood, warm, imbued with spirits, being sent out by the arteries, that which has become cooled and effete is forced on, and all the particles recover their heat which was failing, and their vital stimulus wellnigh exhausted.

Hence it is that if the heart be unaffected, life and health may be restored to almost all the other parts of the body;

but if the heart be chilled, or smitten with any serious disease, it seems matter of necessity that the whole animal fabric should suffer and fall into decay. When the source is corrupted, there is nothing, as Aristotle says,[2] which can be of service either to it or aught that depends on it. And hence, by the way, it may perchance be why grief, and love, and envy, and anxiety, and all affections of the mind of a similar kind are accompanied with emaciation and decay, or with disordered fluids and crudity, which engender all manner of diseases and consume the body of man. For every affection of the mind that is attended with either pain or pleasure, hope or fear, is the cause of an agitation whose influence extends to the heart, and there induces change from the natural constitution, in the temperature, the pulse and the rest, which impairing all nutrition in its source and abating the powers at large, it is no wonder that various forms of incurable disease in the extremities and in the trunk are the consequence, inasmuch as in such circumstances the whole body labours under the effects of vitiated nutrition and a want of native heat.

Moreover, when we see that all animals live through food digested in their interior, it is imperative that the digestion and distribution be perfect, and, as a consequence, that there be a place and receptacle where the aliment is perfected and whence it is distributed to the several members. Now this place is the heart, for it is the only organ in the body which contains blood for the general use; all the others receive it merely for their peculiar or private advantage, just as the heart also has a supply for its own especial behoof in its coronary veins and arteries. But it is of the store which the heart contains in its auricles and ventricles that I here speak. Then the heart is the only organ which is so situated and constituted that it can distribute the blood in due proportion to the several parts of the body, the quantity sent to each being according to the dimensions of the artery which supplies it, the heart serving as a magazine or fountain ready to meet its demands.

[2] De Part. Animal., iii.

Further, a certain impulse or force, as well as an impeller or forcer, such as the heart, was required to effect this distribution and motion of the blood; both because the blood is disposed from slight causes, such as cold, alarm, horror, and the like, to collect in its source, to concentrate like parts to a whole, or the drops of water spilt upon a table to the mass of liquid; and because it is forced from the capillary veins into the smaller ramifications, and from these into the larger trunks by the motion of the extremities and the compression of the muscles generally. The blood is thus more disposed to move from the circumference to the centre than in the opposite direction, even were there no valves to oppose its motion; wherefore, that it may leave its source and enter more confined and colder channels, and flow against the direction to which it spontaneously inclines, the blood requires both force and impelling power. Now such is the heart and the heart alone, and that in the way and manner already explained.

CHAPTER XVI

The Circulation of the Blood is Further Proved from Certain Consequences

There are still certain problems, which, taken as consequences of this truth assumed as proven, are not without their use in exciting belief, as it were, *a posteriore;* and which, although they may seem to be involved in much doubt and obscurity, nevertheless readily admit of having reasons and causes assigned for them. Of such a nature are those that present themselves in connexion with contagions, poisoned wounds, the bites of serpents and rabid animals, lues venerea and the like. We sometimes see the whole system contaminated, though the part first infected remains sound; the lues venerea has occasionally made its attack with pains in the shoulders and head, and other symptoms, the genital organs being all the while unaffected; and then we know that the wound made by a rabid dog having healed, fever and a train of disastrous symptoms may nevertheless supervene. Whence it appears that the contagion impressed upon or deposited in a particular part, is by-and-by carried by the returning current of blood to the heart, and by that organ is sent to contaminate the whole body.

In tertian fever, the morbific cause seeking the heart in the first instance, and hanging about the heart and lungs, renders the patient short-winded, disposed to sighing, and indisposed to exertion, because the vital principle is oppressed and the blood forced into the lungs and rendered thick. It does not pass through them, (as I have myself seen in opening the bodies of those who had died in the beginning of the attack,) when the pulse is always frequent, small, and occasionally irregular; but the heat increasing, the matter becoming attenuated, the passages forced, and the transit made, the whole body begins to rise in temperature, and the pulse becomes fuller and stronger. The febrile paroxysm is fully formed, whilst the preternatural heat kindled in the heart is thence diffused by the arteries through the whole body along with the morbific matter, which is in this way overcome and dissolved by nature.

When we perceive, further, that medicines applied externally exert their influence on the body just as if they had been taken internally, the truth we are contending for is confirmed. Colocynth and aloes in this way move the belly, cantharides excites the urine, garlic applied to the soles of the feet assists expectoration, cordials strengthen, and an infinite number of examples of the same kind might be cited. Perhaps it will not, therefore, be found unreasonable, if we say that the veins, by means of their orifices, absorb some of the things that are applied externally and carry this inwards with the blood, not otherwise, it may be, than those of the mesentery imbibe the chyle from the intestines and carry it mixed with the blood to the liver. For the blood entering the mesentery by the cœliac artery, and the superior and inferior mesenterics, proceeds to the intestines, from which, along with the chyle that has been attracted into the veins, it returns by their numerous ramifications into the vena portæ of the liver, and from this into the vena cava, and this in such wise that the blood in these veins has the same colour and consistency as in other veins, in opposition to what many believe to be the fact. Nor indeed can we imagine two contrary motions in any capillary system—the chyle upwards, the blood downwards. This could scarcely take place, and must be held as altogether improbable. But is not the thing rather arranged as it is by the consummate providence of nature? For were the chyle mingled with the blood, the crude with the digested, in equal proportions, the result would not be concoction, transmutation, and sanguification, but rather, and because they are severally active and passive, a mixture or combination, or medium compound of the two, precisely as happens when wine is mixed with water and syrup. But when a very minute quantity of chyle is mingled with a very large quantity of circulating blood, a quantity of chyle that bears no kind of proportion to the mass of blood, the effect is the same, as Aristotle says, as when a drop of water is added to a cask of wine, or the contrary; the mass does not then present itself as a mixture, but is still sensibly either wine or water.

So in the mesenteric veins of an animal we do not find either chyme or chyle and blood, blended together or distinct, but only blood, the same in colour, consistency, and other sensible properties, as it appears in the veins generally. Still as there is a certain though small and inappreciable portion of chyle or incompletely digested matter mingled with the blood, nature has interposed the liver, in whose meandering channels it suffers delay and undergoes additional change, lest arriving prematurely and crude at the heart, it should oppress the vital principle. Hence in the embryo, there is almost no use for the liver, but the umbilical vein passes directly through, a foramen or an anastomosis existing from the vena portæ. The blood returns from the intestines of the fœtus, not through the liver, but into the umbilical vein mentioned, and flows at once into the heart, mingled with the natural blood which is returning from the placenta; whence also it is that in the development of the fœtus the liver is one of the organs that is last formed. I have observed all the members perfectly marked out in the human fœtus, even the genital organs, whilst there was yet scarcely any trace of the liver. And indeed at the period when all the parts, like the heart itself in the beginning, are still white, and except in the veins there is no appearance of redness, you shall see nothing in the seat of the liver but a shapeless collection, as it were, of extravasated blood, which you might take for the effects of a contusion or ruptured vein.

But in the incubated egg there are, as it were, two umbilical vessels, one from the albumen passing entire through the liver, and going straight to the heart; another from the yelk, ending in the vena portæ; for it appears that the chick, in the first instance, is entirely formed and nourished by the white; but by the yelk after it has come to perfection and is excluded from the shell; for this part may still be found in the abdomen of the chick many days after its exclusion, and is a substitute for the milk to other animals.

But these matters will be better spoken of in my observations on the formation of the fœtus, where many propositions, the following among the number, will be discussed: Wherefore is this part formed or perfected first, that last, and of the several members, what part is the cause of another? And there are many points having special reference to the heart, such as wherefore does it first acquire consistency, and appear to possess life, motion, sense, before any other part of the body is perfected, as Aristotle says in his third book, "De partibus Animalium"? And so also of the blood, wherefore does it precede all the rest? And in what way does it possess the vital and animal principle, and show a tendency to motion, and to be impelled hither and thither, the end for which the heart appears to be made? In the same way, in considering the pulse, why should one kind of pulse indicate death, another recovery? And so of all the other kinds of pulse, what may be the cause and indication of each? Likewise we must consider the reason of crises and natural critical discharges; of nutrition, and especially the distribution of the nutriment; and of defluxions of every description. Finally, reflecting on every part of medicine, physiology, pathology, semeiotics and therapeutics, when I see how many questions can be answered, how many doubts resolved, how much obscurity illustrated by the truth we have declared, the light we have made to shine, I see a field of such vast extent in which I might proceed so far, and expatiate so widely, that this my tractate would not only swell out into a volume, which was beyond my purpose, but my whole life, perchance, would not suffice for its completion.

In this place, therefore, and that indeed in a single chapter, I shall only endeavour to refer the various particulars that present themselves in the dissection of the heart and arteries to their several uses and causes; for so I shall meet with many things which receive light from the truth I have been contending for, and which, in their turn, render it more obvious. And indeed I would have it confirmed and illustrated by anatomical arguments above all others.

There is but a single point which indeed would be more correctly placed among our observations on the use of the spleen, but which it will not be altogether impertinent to notice in this place incidentally. From the splenic branch which passes into the pancreas, and from the upper part, arise the posterior coronary, gastric, and gastroepiploic veins, all of which are distributed upon the stomach in numerous branches and twigs, just as the mesenteric vessels are upon the intestines. In a similar way, from the inferior part of the same splenic branch, and along the back of the colon and rectum proceed the hemorrhoidal veins. The blood returning by these veins, and bringing the cruder juices along with it, on the one hand from the stomach, where they are thin, watery, and not yet perfectly chylified; on the other thick and more earthy, as derived from the fæces, but all poured into this splenic branch, are duly tempered by the admixture of contraries; and nature mingling together these two kinds of juices, difficult of coction by reason of most opposite defects, and then diluting them with a large quantity of warm blood, (for we see that the quantity returned from the spleen must be very large when we contemplate the size of its arteries,) they are brought to the porta of the liver in a state of higher preparation. The defects of either extreme are supplied and compensated by this arrangement of the veins.

CHAPTER XVII

The Motion and Circulation of the Blood are Confirmed from the Particulars Apparent in the Structure of the Heart, and from Those Things which Dissection Unfolds

I do not find the heart as a distinct and separate part in all animals; some, indeed, such as the zoōphytes, have no heart; this is because these animals are coldest, of one great bulk, of soft texture, or of a certain uniform sameness or simplicity of structure; among the number I may instance grubs and earth-worms, and those that are engendered of putrefaction and do not preserve their species. These have no heart, as not requiring any impeller of nourishment into the extreme parts; for they have bodies which are connate and homogeneous and without limbs; so that by the contraction and relaxation of the whole body they assume and expel, move and remove, the aliment. Oysters, mussels, sponges, and the whole genus of zoōphytes or plant-animals have no heart, for the whole body is used as a heart, or the whole animal is a heart. In a great number of animals,—almost the whole tribe of insects—we cannot see distinctly by reason of the smallness of the body; still in bees, flies, hornets, and the like we can perceive something pulsating with the help of a magnifying-glass; in pediculi, also, the same thing may be seen, and as the body is transparent, the passage of the food through the intestines, like a black spot or stain, may be perceived by the aid of the same magnifying-glass.

But in some of the pale-blooded and colder animals, as in snails, whelks, shrimps, and shell-fish, there is a part which pulsates,—a kind of vesicle or auricle without a heart,—slowly, indeed, and not to be perceived except in the warmer season of the year. In these creatures this part is so contrived that it shall pulsate, as there is here a necessity for some impulse to distribute the nutritive fluid, by reason of the variety of organic parts or of the density of the substance; but the pulsations occur unfrequently, and sometimes in consequence of the cold not at all, an arrangement the best adapted to them as being of a doubtful nature, so that sometimes they appear to live, sometimes to die; sometimes they show the vitality of an animal, sometimes of a vegetable. This seems also to be the case with the insects which conceal themselves in winter, and lie, as it were, defunct, or merely manifesting a kind of vegetative existence. But whether the same thing happens in the case of certain animals that have red blood, such as frogs, tortoises, serpents, swallows, may be very properly doubted.

In all the larger and warmer animals which have red blood, there was need of an impeller of the nutritive fluid, and that, perchance, possessing a considerable amount of power. In fishes, serpents, lizards, tortoises, frogs, and others of the same kind there is a heart present, furnished with both an auricle and a ventricle, whence it is perfectly true, as Aristotle has observed,[1] that no sanguineous animal is without a heart, by the impelling power of which the nutritive fluid is forced, both with greater vigour and rapidity, to a greater distance; and not merely agitated by an auricle, as it is in lower forms. And then in regard to animals that are yet larger, warmer, and more perfect, as they abound in blood, which is always hotter and more spirituous, and which possess bodies of greater size and consistency, these require a larger, stronger, and more fleshy heart, in order that the nutritive fluid may be propelled with yet greater force and celerity. And further, inasmuch as the more perfect animals require a still more perfect nutrition, and a larger supply of native heat, in order that the aliment may be thoroughly concocted and acquire the last degree of perfection, they required both lungs and a second ventricle, which should force the nutritive fluid through them.

Every animal that has lungs has, therefore, two ventricles to its heart—one right, the other left; and wherever there is a right, there also is there a left ventricle; but

[1] De Part. Animal., lib. iii.

the contrary of this does not hold good: where there is a left there is not always a right ventricle. The left ventricle I call that which is distinct in office, not in place from the other, that one, namely, which distributes the blood to the body at large, not to the lungs only. Hence the left ventricle seems to form the principle part of the heart; situated in the middle, more strongly marked, and constructed with greater care, the heart seems formed for the sake of the left ventricle, and the right but to minister to it. The right neither reaches to the apex of the heart nor is it nearly of such strength, being three times thinner in its walls, and in some sort jointed on to the left (as Aristotle says), though, indeed, it is of greater capacity, inasmuch as it has not only to supply material to the left ventricle, but likewise to furnish aliment to the lungs.

It is to be observed, however, that all this is otherwise in the embryo, where there is not such a difference between the two ventricles. There, as in a double nut, they are nearly equal in all respects, the apex of the right reaching to the apex of the left, so that the heart presents itself as a sort of double-pointed cone. And this is so, because in the fœtus, as already said, whilst the blood is not passing through the lungs from the right to the left cavities of the heart, it flows by the foramen ovale and ductus arteriosus directly from the vena cava into the aorta, whence it is distributed to the whole body. Both ventricles have, therefore, the same office to perform, whence their equality of constitution. It is only when the lungs come to be used and it is requisite that the passages indicated should be blocked up that the difference in point of strength and other things between the two ventricles begins to be apparent. In the altered circumstances the right has only to drive the blood through the lungs, whilst the left has to propel it through the whole body.

There are, moreover, within the heart numerous braces, in the form of fleshy columns and fibrous bands, which Aristotle, in his third book on "Respiration," and the "Parts of Animals," entitles nerves. These are variously extended, and are either distinct or contained in grooves in the walls and partition, where they occasion numerous pits or depressions. They constitute a kind of small muscles, which are superadded and supplementary to the heart, assisting it to execute a more powerful and perfect contraction, and so proving subservient to the complete expulsion of the blood. They are, in some sort, like the elaborate and artful arrangement of ropes in a ship, bracing the heart on every side as it contracts, and so enabling it more effectually and forcibly to expel the charge of blood from its ventricles. This much is plain, at all events, that in some animals they are less strongly marked than in others; and, in all that have them, they are more numerous and stronger in the left than in the right ventricle; and while some have them present in the left, yet they are absent in the right ventricle. In man they are more numerous in the left than in the right ventricle, more abundant in the ventricles than in the auricles; and occasionally there appear to be none present in the auricles. They are numerous in the large, more muscular and hardier bodies of countrymen, but fewer in more slender frames and in females.

In those animals in which the ventricles of the heart are smooth within and entirely without fibres of muscular bands, or anything like hollow pits, as in almost all the smaller birds, the partridge and the common fowl, serpents, frogs, tortoises, and most fishes, there are no chordæ tendineæ, nor bundles of fibres, neither are there any tricuspid valves in the ventricles.

Some animals have the right ventricle smooth internally, but the left provided with fibrous bands, such as the goose, swan, and larger birds; and the reason is the same here as elsewhere. As the lungs are spongy and loose and soft, no great amount of force is required to force the blood through them; therefore the right ventricle is either without the bundles in question, or they are fewer and weaker, and not so fleshy or like muscles. Those of the left ventricle, however, are both stronger and more numerous, more fleshy and muscular, because the left ventricle requires to be stronger, inasmuch as the blood which it propels has to be driven through the whole body. And this, too, is the reason why the left ventricle occupies the middle of the heart, and has parietes three times thicker and stronger than those of the right. Hence all animals—and among men it is similar—

that are endowed with particularly strong frames, and with large and fleshy limbs at a great distance from the heart, have this central organ of greater thickness, strength, and muscularity. This is manifest and necessary. Those, on the contrary, that are of softer and more slender make have the heart more flaccid, softer, and internally either less or not at all fibrous. Consider, farther, the use of the several valves, which are all so arranged that the blood, once received into the ventricles of the heart, shall never regurgitate; once forced into the pulmonary artery and aorta, shall not flow back upon the ventricles. When the valves are raised and brought together, they form a three-cornered line, such as is left by the bite of a leech; and the more they are forced, the more firmly do they oppose the passage of the blood. The tricuspid valves are placed, like gate-keepers, at the entrance into the ventricles from the venæ cavæ and pulmonary veins, lest the blood when most forcibly impelled should flow back. It is for this reason that they are not found in all animals, nor do they appear to have been constructed with equal care in all animals in which they are found. In some they are more accurately fitted, in others more remissly or carelessly contrived, and always with a view to their being closed under a greater or a slighter force of the ventricle. In the left ventricle, therefore, in order that the occlusion may be the more perfect against the greater impulse, there are only two valves, like a mitre, and produced into an elongated cone, so that they come together and touch to their middle; a circumstance which perhaps led Aristotle into the error of supposing this ventricle to be double, the division taking place transversely. For the same reason, and that the blood may not regurgitate upon the pulmonary veins, and thus the force of the ventricle in propelling the blood through the system at large come to be neutralized, it is that these mitral valves excel those of the right ventricle in size and strength and exactness of closing. Hence it is essential that there can be no heart without a ventricle, since this must be the source and store-house of the blood. The same law does not hold good in reference to the brain. For almost no genus of birds has a ventricle in the brain, as is obvious in the goose and swan, the brains of which nearly equal that of a rabbit in size; now rabbits have ventricles in the brain, whilst the goose has none. In like manner, wherever the heart has a single ventricle, there is an auricle appended, flaccid, membranous, hollow, filled with blood; and where there are two ventricles, there are likewise two auricles. On the other hand, some animals have an auricle without any ventricle; or, at all events, they have a sac analogous to an auricle; or the vein itself, dilated at a particular part, performs pulsations, as is seen in hornets, bees, and other insects, which certain experiments of my own enable me to demonstrate, have not only a pulse, but a respiration in that part which is called the tail, whence it is that this part is elongated and contracted now more rarely, now more frequently, as the creature appears to be blown and to require a large quantity of air. But of these things, more in our "Treatise on Respiration."

It is in like manner evident that the auricles pulsate, contract, as I have said before, and throw the blood into the ventricles; so that wherever there is a ventricle, an auricle is necessary, not merely that it may serve, according to the general belief, as a source and magazine for the blood: for what were the use of its pulsations had it only to contain?

The auricles are prime movers of the blood, especially the right auricle, which, as already said, is "the first to live, the last to die"; whence they are subservient to sending the blood into the ventricles, which, contracting continuously, more readily and forcibly expel the blood already in motion; just as the ball-player can strike the ball more forcibly and further if he takes it on the rebound than if he simply threw it. Moreover, and contrary to the general opinion, neither the heart nor anything else can dilate or distend itself so as to draw anything into its cavity during the diastole, unless, like a sponge, it has been first compressed and is returning to its primary condition. But in animals all local motion proceeds from, and has its origin in, the contraction of some part; consequently it is by the contraction of the auricles that the blood is thrown into the ventricles, as I have already shown, and from there, by the contraction of the ventricles, it is propelled and distributed. Concerning local motions, it is true that the immediate moving organ in every motion of an animal primarily endowed with a motive spirit (as Aristotle has it[2]) is contractile; in which way the word νεῦρου is derived from νεύω, nuto, contraho; and if I am permitted to proceed in my purpose of making a particular demonstration of the organs of motion in animals from observations in my possession, I trust I shall be able to make sufficiently plain how Aristotle was acquainted with the muscles, and advisedly referred all motion in animals to the nerves, or to the contractile element, and, therefore, called those little bands in the heart nerves.

But that we may proceed with the subject which we have in hand, viz., the use of the auricles in filling the ventricles, we should expect that the more dense and compact the heart, the thicker its parietes, the stronger and more muscular must be the auricle to force and fill it, and vice versâ. Now this is actually so: in some the auricle presents itself as a sanguinolent vesicle, as a thin membrane containing blood, as in fishes, in which the sac that stands in lieu of the auricles is of such delicacy and ample capacity that it seems to be suspended or to float above the heart. In those fishes in which the sac is somewhat more fleshy, as in the carp, barbel, tench, and others, it bears a wonderful and strong resemblance to the lungs.

In some men of sturdier frame and stouter make the right auricle is so strong, and so curiously constructed on its inner surface of bands and variously interlacing fibres, that it seems to equal in strength the ventricle of the heart in other subjects; and I must say that I am astonished to find such diversity in this particular in different individuals. It is to be observed, however, that in the fœtus the auricles are out of all proportion large, which is because they are present before the heart makes its appearance or suffices for its office even when it has appeared, and they, therefore, have, as it were, the duty of the whole heart committed to them, as has already been demonstrated. But what I have observed in the formation of the fœtus, as before remarked (and Aristotle had already confirmed all in studying the incubated egg), throws the greatest light and likelihood upon the point. Whilst the fœtus is yet in the form of a soft worm, or, as is commonly said, in the milk, there is a mere bloody

[2] In the book de Spiritu, and elsewhere.

point or pulsating vesicle, a portion apparently of the umbilical vein, dilated at its commencement or base. Afterwards, when the outline of the fœtus is distinctly indicated and it begins to have greater bodily consistence, the vesicle in question becomes more fleshy and stronger, changes its position, and passes into the auricles, above which the body of the heart begins to sprout, though as yet it apparently performs no office. When the fœtus is farther advanced, when the bones can be distinguished from the fleshy parts and movements take place, then it also has a heart which pulsates, and, as I have said, throws blood by either ventricle from the vena cava into the arteries.

Thus nature, ever perfect and divine, doing nothing in vain, has neither given a heart where it was not required, nor produced it before its office had become necessary; but by the same stages in the development of every animal, passing through the forms of all, as I may say (ovum, worm, fœtus), it acquires perfection in each. These points will be found elsewhere confirmed by numerous observations on the formation of the fœtus.

Finally, it is not without good grounds that Hippocrates in his book, "De Corde," entitles it a muscle; its action is the same; so is its functions, viz., to contract and move something else—in this case the charge of the blood.

Farther, we can infer the action and use of the heart from the arrangement of its fibres and its general structures, as in muscles generally. All anatomists admit with Galen that the body of the heart is made up of various courses of fibres running straight, obliquely, and transversely, with refference to one another; but in a heart which has been boiled, the arrangement of the fibres is seen to be different. All the fibres in the parietes and septum are circular, as in the sphincters; those, again, which are in the columns extend lengthwise, and are oblique longitudinally; and so it comes to pass that when all the fibres contract simultaneously, the apex of the cone is pulled towards its base by the columns, the walls are drawn circularly together into a globe—the whole heart, in short, is contracted and the ventricles narrowed. It is, therefore, impossible not to perceive that, as the action of the organ is so plainly contraction, its function is to propel the blood into the arteries.

Nor are we the less to agree with Aristotle in regard to the importance of the heart, or to question if it receives sense and motion from the brain, blood from the liver, or whether it be the origin of the veins and of the blood, and such like. They who affirm these propositions overlook, or do not rightly understand, the principal argument, to the effect that the heart is the first part which exists, and that it contains within itself blood, life, sensation, and motion, before either the brain or the liver were created or had appeared distinctly, or, at all events, before they could perform any function. The heart, ready furnished with its proper organs of motion, like a kind of internal creature, existed before the body. The first to be formed, nature willed that it should afterwards fashion, nourish, preserve, complete the entire animal, as its work and dwelling-place: and as the prince in a kingdom, in whose hands lie the chief and highest authority, rules over all, the heart is the source and foundation from which all power is derived, on which all power depends in the animal body.

Many things having reference to the arteries farther illustrate and confirm this truth. Why does not the pulmonary vein pulsate, seeing that it is numbered among the arteries? Or wherefore is there a pulse in the pulmonary artery? Because the pulse of the arteries is derived from the impulse of the blood. Why does an artery differ so much from a vein in the thickness and strength of its coats? Because it sustains the shock of the impelling heart and streaming blood. Hence, as perfect nature does nothing in vain, and suffices under all circumstances, we find that the nearer the arteries are to the heart, the more do they differ from the veins in structure; here they are both stronger and more ligamentous, whilst in extreme parts of the body, such as the feet and hands, the brain, the mesentery, and the testicles, the two orders of vessels are so much alike that it is impossible to distinguish between them with the eye. Now this is for the following very sufficient reasons: the more remote the vessels are from the heart, with so much the less force are they distended by the stroke of the heart, which is broken by the great distance at which it is given. Add to this that the impulse of the heart exerted upon the mass of blood, which must needs fill the trunks and branches of the arteries, is diverted, divided, as it were, and diminished at every subdivision, so that the ultimate capillary divisions of the arteries look like veins, and this not merely in constitution, but in function. They have either no perceptible pulse, or they rarely exhibit one, and never except where the heart beats more violently than usual, or at a part where the minute vessel is more dilated or open than elsewhere. It, therefore, happens that at times we are aware of a pulse in the teeth, in inflammatory tumours, and in the fingers; at another time we feel nothing of the sort. By this single symptom I have ascertained for certain that young persons whose pulses are naturally rapid were labouring under fever; and in like manner, on compressing the fingers in youthful and delicate subjects during a febrile paroxysm, I have readily perceived the pulse there. On the other hand, when the heart pulsates more languidly, it is often impossible to feel the pulse not merely in the fingers, but the wrist, and even at the temple, as in persons afflicted with lipothymiæ asphyxia, or hysterical symptoms, and in the debilitated and moribund.

Here surgeons are to be advised that, when the blood escapes with force in the amputation of limbs, in the removal of tumours, and in wounds, it constantly comes from an artery; not always indeed per saltum, because the smaller arteries do not pulsate, especially if a tourniquet has been applied.

For the same reason the pulmonary artery not only has the structure of an artery, but it does not differ so widely from the veins in the thickness of its walls as does the aorta. The aorta sustains a more powerful shock from the left than the pulmonary artery does from the right ventricle, and the walls of this last vessel are thinner and softer than those of the aorta in the same proportion as the walls of the right ventricle of the heart are weaker and thinner than those of the left ventricle. In like manner the lungs are softer and laxer in structure than the flesh and other constituents of the body, and in a similar way the walls of the branches of the pulmonary artery differ from those of the vessels derived from the aorta. And the same proportion in these particulars is universally preserved. The more muscular and powerful men are, the firmer their flesh; the stronger, thicker, denser, and more fibrous their hearts, the thicker, closer, and stronger are the auricles and arteries. Again, in those animals the ventricles of whose hearts are smooth on their inner surface, without villi or valves, and the walls of which are thin, as in fishes, serpents, birds, and very many genera of animals, the arteries differ little or nothing in the thickness of their coats from the veins.

Moreover, the reason why the lungs have such ample vessels, both arteries and veins (for the capacity of the pulmonary veins exceeds that of both crural and jugular vessels), and why they contain so large a quantity of blood, as by experience and ocular inspection we know they do, admonished of the fact indeed by Aristotle, and not led into error by the appearances found in animals which have been bled to death, is, because the blood has its fountain, and storehouse, and the workshop of its last perfection, in the heart and lungs. Why, in the same way, we find in the course of our anatomical dissections the pulmonary vein and left ventricle so full of blood, of the same black colour and clotted character as that with which the right ventricle and pulmonary artery are filled, is because the blood is incessantly passing from one side of the heart to the other through the lungs. Wherefore, in fine, the pulmonary artery has the structure of an artery, and the pulmonary veins have the structure of veins. In function and constitution and everything else the first is an artery, the others are veins, contrary to what is commonly believed; and the reason why the pulmonary artery has so large an orifice is because it transports much more blood than is requisite for the nutrition of the lungs.

All these appearances, and many others, to be noted in the course of dissection, if rightly weighed, seem clearly to illustrate and fully to confirm the truth contended for throughout these pages, and at the same time to oppose the vulgar opinion; for it would be very difficult to explain in any other way to what purpose all is constructed and arranged as we have seen it to be.

GREAT BOOKS IN PHILOSOPHY PAPERBACK SERIES

ESTHETICS

- ❑ Aristotle—*The Poetics*
- ❑ Aristotle—*Treatise on Rhetoric*

ETHICS

- ❑ Aristotle—*The Nicomachean Ethics*
- ❑ Marcus Aurelius—*Meditations*
- ❑ Jeremy Bentham—*The Principles of Morals and Legislation*
- ❑ John Dewey—*Human Nature and Conduct*
- ❑ John Dewey—*The Moral Writings of John Dewey, Revised Edition*
- ❑ Epictetus—*Enchiridion*
- ❑ David Hume—*An Enquiry Concerning the Principles of Morals*
- ❑ Immanuel Kant—*Fundamental Principles of the Metaphysic of Morals*
- ❑ John Stuart Mill—*Utilitarianism*
- ❑ George Edward Moore—*Principia Ethica*
- ❑ Friedrich Nietzsche—*Beyond Good and Evil*
- ❑ Plato—*Protagoras, Philebus, and Gorgias*
- ❑ Bertrand Russell—*Bertrand Russell On Ethics, Sex, and Marriage*
- ❑ Arthur Schopenhauer—*The Wisdom of Life* and *Counsels and Maxims*
- ❑ Adam Smith—*The Theory of Moral Sentiments*
- ❑ Benedict de Spinoza—*Ethics* and *The Improvement of the Understanding*

LOGIC

- ❑ George Boole—*The Laws of Thought*

METAPHYSICS/EPISTEMOLOGY

- ❑ Aristotle—*De Anima*
- ❑ Aristotle—*The Metaphysics*
- ❑ Francis Bacon—*Essays*
- ❑ George Berkeley—*Three Dialogues Between Hylas and Philonous*
- ❑ W. K. Clifford—*The Ethics of Belief and Other Essays*
- ❑ René Descartes—*Discourse on Method* and *The Meditations*
- ❑ John Dewey—*How We Think*
- ❑ John Dewey—*The Influence of Darwin on Philosophy and Other Essays*
- ❑ Epicurus—*The Essential Epicurus: Letters, Principal Doctrines, Vatican Sayings, and Fragments*
- ❑ Sidney Hook—*The Quest for Being*
- ❑ David Hume—*An Enquiry Concerning Human Understanding*
- ❑ David Hume—*A Treatise on Human Nature*
- ❑ William James—*The Meaning of Truth*
- ❑ William James—*Pragmatism*
- ❑ Immanuel Kant—*The Critique of Judgment*
- ❑ Immanuel Kant—*Critique of Practical Reason*
- ❑ Immanuel Kant—*Critique of Pure Reason*
- ❑ Gottfried Wilhelm Leibniz—*Discourse on Metaphysics* and *The Monadology*
- ❑ John Locke—*An Essay Concerning Human Understanding*
- ❑ George Herbert Mead—*The Philosophy of the Present*

- ❑ Charles S. Peirce—*The Essential Writings*
- ❑ Plato—*The Euthyphro*, *Apology*, *Crito*, and *Phaedo*
- ❑ Plato—*Lysis*, *Phaedrus*, *and Symposium*
- ❑ Bertrand Russell—*The Problems of Philosophy*
- ❑ George Santayana—*The Life of Reason*
- ❑ Sextus Empiricus—*Outlines of Pyrrhonism*
- ❑ Ludwig Wittgenstein—*Wittgenstein's Lectures: Cambridge, 1932–1935*

PHILOSOPHY OF RELIGION

- ❑ Jeremy Bentham—*The Influence of Natural Religion on the Temporal Happiness of Mankind*
- ❑ Marcus Tullius Cicero—*The Nature of the Gods* and *On Divination*
- ❑ Ludwig Feuerbach—*The Essence of Christianity*
- ❑ Paul Henri Thiry, Baron d'Holbach—*Good Sense*
- ❑ David Hume—*Dialogues Concerning Natural Religion*
- ❑ William James—*The Varieties of Religious Experience*
- ❑ John Locke—*A Letter Concerning Toleration*
- ❑ Lucretius—*On the Nature of Things*
- ❑ John Stuart Mill—*Three Essays on Religion*
- ❑ Friedrich Nietzsche—*The Antichrist*
- ❑ Thomas Paine—*The Age of Reason*
- ❑ Bertrand Russell—*Bertrand Russell On God and Religion*

SOCIAL AND POLITICAL PHILOSOPHY

- ❑ Aristotle—*The Politics*
- ❑ Mikhail Bakunin—*The Basic Bakunin: Writings, 1869–1871*
- ❑ Edmund Burke—*Reflections on the Revolution in France*
- ❑ John Dewey—*Freedom and Culture*
- ❑ John Dewey—*Individualism Old and New*
- ❑ John Dewey—*Liberalism and Social Action*
- ❑ G. W. F. Hegel—*The Philosophy of History*
- ❑ G. W. F. Hegel—*Philosophy of Right*
- ❑ Thomas Hobbes—*The Leviathan*
- ❑ Sidney Hook—*Paradoxes of Freedom*
- ❑ Sidney Hook—*Reason, Social Myths, and Democracy*
- ❑ John Locke—*The Second Treatise on Civil Government*
- ❑ Niccolo Machiavelli—*The Prince*
- ❑ Karl Marx (with Friedrich Engels)—*The Economic and Philosophic Manuscripts of 1844* and *The Communist Manifesto*
- ❑ Karl Marx (with Friedrich Engels)—*The German Ideology*, including *Theses on Feuerbach* and *Introduction to the Critique of Political Economy*
- ❑ Karl Marx—*The Poverty of Philosophy*
- ❑ John Stuart Mill—*Considerations on Representative Government*
- ❑ John Stuart Mill—*On Liberty*
- ❑ John Stuart Mill—*On Socialism*
- ❑ John Stuart Mill—*The Subjection of Women*
- ❑ Montesquieu, Charles de Secondat—*The Spirit of Laws*
- ❑ Friedrich Nietzsche—*Thus Spake Zarathustra*

- ❑ Thomas Paine—*Common Sense*
- ❑ Thomas Paine—*Rights of Man*
- ❑ Plato—*Laws*
- ❑ Plato—*The Republic*
- ❑ Jean-Jacques Rousseau—*Émile*
- ❑ Jean-Jacques Rousseau—*The Social Contract*
- ❑ Mary Wollstonecraft—*A Vindication of the Rights of Men*
- ❑ Mary Wollstonecraft—*A Vindication of the Rights of Women*

GREAT MINDS PAPERBACK SERIES

ART

- ❑ Leonardo da Vinci—*A Treatise on Painting*

ECONOMICS

- ❑ Charlotte Perkins Gilman—*Women and Economics: A Study of the Economic Relation between Women and Men*
- ❑ John Maynard Keynes—*The General Theory of Employment, Interest, and Money*
- ❑ John Maynard Keynes—*A Tract on Monetary Reform*
- ❑ Thomas R. Malthus—*An Essay on the Principle of Population*
- ❑ Alfred Marshall—*Money, Credit, and Commerce*
- ❑ Alfred Marshall—*Principles of Economics*
- ❑ Karl Marx—*Theories of Surplus Value*
- ❑ John Stuart Mill—*Principles of Political Economy*
- ❑ David Ricardo—*Principles of Political Economy and Taxation*
- ❑ Adam Smith—*Wealth of Nations*
- ❑ Thorstein Veblen—*Theory of the Leisure Class*

HISTORY

- ❑ Edward Gibbon—*On Christianity*
- ❑ Alexander Hamilton, John Jay, and James Madison—*The Federalist*
- ❑ Herodotus—*The History*
- ❑ Charles Mackay—*Extraordinary Popular Delusions and the Madness of Crowds*
- ❑ Thucydides—*History of the Peloponnesian War*

LAW

- ❑ John Austin—*The Province of Jurisprudence Determined*

LITERATURE

- ❑ Jonathan Swift—*A Modest Proposal and Other Satires*
- ❑ H. G. Wells—*The Conquest of Time*

PSYCHOLOGY

- ❑ Sigmund Freud—*Totem and Taboo*

RELIGION/FREETHOUGHT

- ❑ Desiderius Erasmus—*The Praise of Folly*
- ❑ Thomas Henry Huxley—*Agnosticism and Christianity and Other Essays*
- ❑ Ernest Renan—*The Life of Jesus*
- ❑ Upton Sinclair—*The Profits of Religion*
- ❑ Elizabeth Cady Stanton—*The Woman's Bible*
- ❑ Voltaire—*A Treatise on Toleration and Other Essays*
- ❑ Andrew D. White—*A History of the Warfare of Science with Theology in Christendom*

SCIENCE

- ❑ Jacob Bronowski—*The Identity of Man*
- ❑ Nicolaus Copernicus—*On the Revolutions of Heavenly Spheres*
- ❑ Marie Curie—*Radioactive Substances*
- ❑ Charles Darwin—*The Autobiography of Charles Darwin*
- ❑ Charles Darwin—*The Descent of Man*
- ❑ Charles Darwin—*The Origin of Species*
- ❑ Charles Darwin—*The Voyage of the* Beagle
- ❑ René Descartes—*Treatise of Man*
- ❑ Albert Einstein—*Relativity*
- ❑ Michael Faraday—*The Forces of Matter*
- ❑ Galileo Galilei—*Dialogues Concerning Two New Sciences*
- ❑ Ernst Haeckel—*The Riddle of the Universe*
- ❑ William Harvey—*On the Motion of the Heart and Blood in Animals*
- ❑ Werner Heisenberg—*Physics and Philosophy*
- ❑ Julian Huxley—*Evolutionary Humanism*
- ❑ Thomas H. Huxley—*Evolution and Ethics* and *Science and Morals*
- ❑ Edward Jenner—*Vaccination against Smallpox*
- ❑ Johannes Kepler—*Epitome of Copernican Astronomy* and *Harmonies of the World*
- ❑ James Clerk Maxwell—*Matter and Motion*
- ❑ Isaac Newton—*Opticks, Or Treatise of the Reflections, Inflections, and Colours of Light*
- ❑ Isaac Newton—*The Principia*
- ❑ Louis Pasteur and Joseph Lister—*Germ Theory and Its Application to Medicine* and *On the Antiseptic Principle of the Practice of Surgery*
- ❑ William Thomson (Lord Kelvin) and Peter Guthrie Tait—*The Elements of Natural Philosophy*
- ❑ Alfred Russel Wallace—*Island Life*

SOCIOLOGY

- ❑ Emile Durkheim—*Ethics and the Sociology of Morals*